53 **Topics in Current Chemistry**
Fortschritte der chemischen Forschung

Gas-Phase Electron Diffraction

Springer-Verlag
Berlin Heidelberg GmbH 1975

This series presents critical reviews of the present position and future trends in modern chemical research. It is addressed to all research and industrial chemists who wish to keep abreast of advances in their subject.

As a rule, contributions are specially commissioned. The editors and publishers will, however, always be pleased to receive suggestions and supplementary information. Papers are accepted for "Topics in Current Chemistry" in either German or English.

ISBN 978-3-662-15865-4 ISBN 978-3-540-37326-1 (eBook)
DOI 10.1007/978-3-540-37326-1

Originally published by Springer-Verlag Berlin Heidelberg New York in 1975
Softcover reprint of the hardcover 1st edition 1975

Library of Congress Cataloging in Publication Data. Main entry under title: Gas-phase electron diffraction. (Topics in current chemistry: 53) Bibliography: p. Includes index. 1. Electrons–Diffraction–Addresses, essays,lectures. 2. Organometallic compounds–Analysis–Addresses, essays, lectures. I. Haaland, A. II. Series. QD1. F58 vol. 53 [QD117. E4] 540'. 8s [539.7'2112]. 74-28208.
Typesetting and printing: Schwetzinger Verlagsdruckerei GmbH, Schwetzingen. Bookbinding: Konrad Triltsch, Graphischer Betrieb, Würzburg.

Contents

Organometallic Compounds Studied by Gas-Phase Electron Diffraction

Arne Haaland 1

Stereochemistry of Compounds Containing Bonds between Si, P, S, Cl, and N or O

Lev Vilkov and Leonid S. Khaikin 25

The Geometric and Dynamic Structures of Fluorocarbons and Related Compounds

Akimichi Yokozeki and Simon H. Bauer 71

Springer-Verlag — D-6900 Heidelberg 1 · Postfach 105 280
Telephone (0 62 21) 48 7-1 · Telex 04-61723
D-1000 Berlin 33 · Heidelberger Platz 3
Telephone (0 30) 82 20 11 · Telex 01-833 19

Springer-Verlag New York Inc. — New York, NY 10010 · 175, Fifth Avenue
Telephone 673-26 60

Organometallic Compounds Studied by Gas-Phase Electron Diffraction

Dr. Arne Haaland

Department of Chemistry, University of Oslo, Blindern, Oslo, Norway

Contents

1. Introduction 2
2. Beryllium 2
3. Magnesium 6
4. Boron 7
5. Aluminum 8
6. Cyclopentadienyl Derivatives of other Main-Group Elements 17
7. Ferrocene and other Biscyclopentadienyl Derivatives of the Transition Elements . 19
8. References 21

1. Introduction

It is not the purpose of this article to provide an introduction to the principles and techniques of gas-phase electron diffraction as a tool for the determination of molecular structure. This task has already been accomplished in a recent book[1] and several recent review articles[2–6]. Neither is it my aim to review all the organometallic structures determined by gas-phase electron diffraction, since this too has been done recently[7]. I propose to concentrate on a few areas of organometallic chemistry where a number of related compounds have been studied by gas-phase electron diffraction and where patterns are becoming apparent. These patterns are discussed in terms of simple molecular orbital theory or the valence-shell repulsion theory[8], or in terms of non-bonded repulsions. I have omitted one area where electron diffraction, mainly through the work of British and Russian scientists, has made large contributions, namely the organometallic chemistry of the main group IV elements. This area is, at least in part, covered in the article by Dr. Vilkov in this volume.

It is hardly necessary to stress the usefulness of electron diffraction as a tool in structure chemistry. It should be sufficient to point out that at least eight of the species discussed here are inaccessible to X-ray diffraction since these compounds change their degree of association on going into the solid phase.

Throughout this article internuclear distances are given as r_a[6] and estimated standard deviations (when stated in the original work) are given in parentheses in units of the last digit.

2. Beryllium[a]

At room temperature dimethylberyllium forms colorless crystals with very low vapor pressure. An X-ray diffraction study by Snow and Rundle[12] showed the crystal to consist of infinite straight chains of beryllium atoms bridged by methyl groups (Fig. 1). This was the first case in which bridging methyl groups were established by X-ray diffraction[b].

The Be–C bridge bonds are considerably longer than the single Be–C bonds in monomeric dimethylberyllium or di-*tert*-butylberyllium = 1.70 Å (see below), and 0.10 Å longer than the sum of the tetrahedral covalent radii[16] of Be and C = 1.83 Å. The Be–Be distance on the other hand is slightly *shorter* than twice the tetrahedral covalent radius of Be = 2.12 Å. Insofar as bond distances in "electron-deficient" molecules of this type can give any information about bond strength, they indicate that Be–Be bonding is as strong as Be–C bonding. Indeed, *ab-initio* molecular orbital calculations on the methyl-bridged dimer $(HBeCH_3)_2$[17] yield a Be–Be overlap population of 0.32 that is very similar to the Be–C bridge overlap population = 0.35. We shall see below that similar conditions prevail in dimeric trimethylaluminum, in diborane, and in dimeric dimethylaluminum hydride.

The methyl groups in polymeric dimethylberyllium are very tightly packed: the distance between C atoms in neighboring rings is 3.11 Å compared to a van der Waals diameter of 4.0 Å for a methyl group[16]. If the bridging methyl groups are displaced by more bulky

---

a) For recent reviews of organoberyllium chemistry, see Refs. [9–11].

b) The methyl-bridged "tetramethylplatinum tetramer" described earlier[13] was in fact trimethylplatinum hydroxide[14, 15].

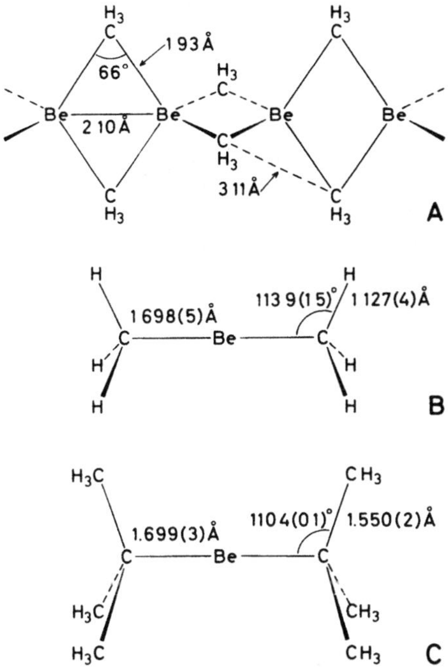

Fig. 1. The structure of dimethylberyllium in the crystalline[12] (*A*) and in the gas phase[20] (*B*) and of di-*tert*-butylberyllium in the gas phase[19] (*C*)

alkyl groups, the van der Waals strain increases. This is probably the reason why diethylberyllium becomes liquid at room temperature. The compound is dimeric in hydrocarbon solution, but the pure liquid consists of more highly associated species. Di-isopropylberyllium is dimeric in hydrocarbon solution, while di-*tert*-butylberyllium is monomeric.

Infrared and Raman spectra show that the C–Be–C skeleton of di-*tert*-butylberyllium is linear and indicate that the barrier to internal rotation is low[18]. The bond distances and valence angles determined by gas-phase electron diffraction[19] are shown in Fig. 1. The electron diffraction data are not consistent with models of D_{3d} symmetry (staggered *tert*-butyl groups) but are consistent with models of D_{3h} symmetry with eclipsed or freely rotating *tert*-butyl groups. There is no van der Waals contact between the two *tert*-butyl groups, and it seems reasonable to assume that the barrier to internal rotation amounts to no more than a few calories.

Kovar and Morgan[20] have recorded the infrared absorption spectra of gaseous dimethylberyllium over the temperature range 125° to 180 °C. The spectra of saturated vapors contained a few lines that were attributed to associated species (probably dimeric), but the unsaturated vapors seemed to consist of monomeric species only, the effective symmetry of the monomer being D_{3h}. This implies that the C–Be–C skeleton is linear and that the barrier to internal rotation of the methyl groups is negligible. The asymmetric C–Be–C stretching frequency was assigned at ν_7 = 1081 cm^{-1}. In di-*tert*-butylberyllium the corresponding vibration had been assigned at much lower frequency, ν_6 = 458 cm^{-1}. Kovar and Morgan therefore suggested that the Be–C bonds in monomeric dimethylberyllium were significantly strengthened by hyperconjugation, *i.e.* the C–H bonding electrons were partly delocalized into the empty $2p\pi$ atomic orbitals on Be.

The electron diffraction pattern from gaseous dimethylberyllium has been recorded at a reservoir temperature of about 150 °C and a nozzle temperature of about 160 °C[21]. It was found that the amount of associated species in the gas jet must have been negligible. The bond distances and valence angles obtained for the monomer are shown in Fig. 1.

During least-squares refinement of the structure the C $\cdots$ C distance was refined as an independent parameter. The value obtained was 0.035(13) Å less than twice the Be–C bond distance. This difference can be explained either by the assumption that the $<$ C–Be–C equilibrium angle is 163° rather than 180°, or as the result of shrinkage, *i.e.* because large-amplitude bending vibrations make the $<$ C–Be–C angle on average less than 180° (though the equilibrium angle is not). We prefer the second explanation. The C–Be–C bending frequency has not been assigned, but since there are no lone-pair electrons on Be to resist deformation, it may be very low.

The contribution of the long H $\cdots$ H distances to the molecular intensity was too small to allow the relative orientation of the methyl groups to be determined.

The Be–C bond distance in dimethylberyllium is indistinguishable from the Be–C bond distance in di-*tert*-butylberyllium, though a difference of the order of 0.015 Å or less cannot be ruled out. The observed bond distances therefore offer no support for the theory of hyperconjugation.

Baird and co-workers have published the results of *ab-initio* molecular orbital calculations on $HBeCH_3$[17]. First the Be–H and Be–C bond distances were varied to minimize the energy with a full $2p$ basis set on Be. The bond distances were found to be 1.33 Å and 1.71 Å, respectively. The latter is in good agreement with the Be–C bond distance in dimethylberyllium. Removal of the $2p\pi$ orbitals on Be from the basis set resulted in an increase of 7.1 kcal mole^{-1} in the energy and 0.05 Å in the Be–C distance. One wonders whether calculations on di-*tert*-butylberyllium would yield similar results, and whether in that case hyperconjugation is a useful concept.

Dimethylberyllium forms both 1 : 1 and 1 : 2 complexes with tertiary amines. The 1 : 1 complex with trimethylamine is reported to be stable in the gas phase, but an attempt to determine the molecular structure by gas-phase electron diffraction was unseccessful. The scattering pattern recorded at a reservoir temperature of 65 °C and a nozzle temperature of 85 °C showed that the gas jet must have consisted of nearly 100% trimethylamine[22]. Clearly, the complex had dissociated and the dimethylberyllium had condensed on the walls of the inlet system.

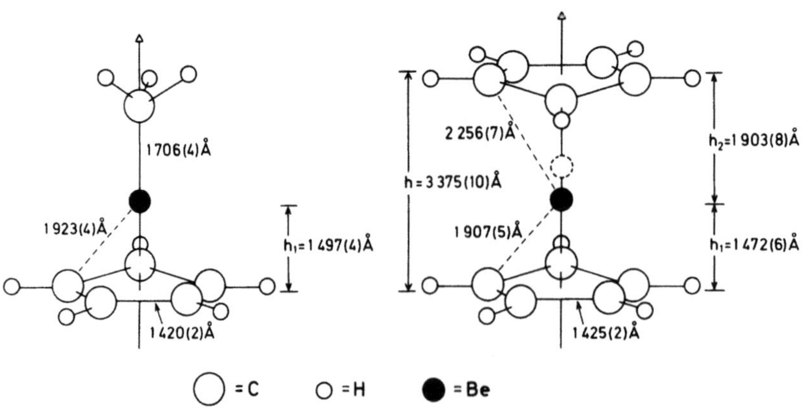

Fig. 2. The molecular structures of methyl(cyclopentadienyl)beryllium[24] and biscyclopentadienylberyllium[27, 28] in the gas phase

The structure of the 1 : 2 complex with quinuclidine has, however, been determined by X-ray diffraction[23]. The coordination about Be is distorted tetrahedral, the < C–Be–C angle being 118°, and < N–Be–N 111°. The average Be–C bond distance is 1.83 Å, *i.e.* 0.13 Å longer than in free dimethylberyllium, and the average Be–N distance is 1.91 Å.

The increase of the Be–C bond distance in the complex as well as the relative magnitudes of the < C–Be–C and < N–Be–N angles is easily rationalized in terms of the valence-shell electron-pair repulsion (VSEPR) theory[8]. In the complex, repulsion between the electron pairs partly introduced into the formerly empty $2p$ orbitals of dimethylberyllium and the electron pairs in the Be–C bonds would be expected to weaken the Be–C bonds and push them together.

Morgan and co-workers have synthesized a number of monocyclopentadienyl derivatives of beryllium, and several of them have been investigated by gas-phase electron diffraction. All have been found to contain *penta-hapto* cyclopentadienyl, h^5–Cp or π–Cp rings. The structure of methyl(cyclopentadienyl)beryllium[24] is shown in Fig. 2[c]. The Be atom is situated on the C_5 symmetry axis, 1.50 Å above the plane of the ring. The Be–C(Cp) bond distance is similar to the Be–C bridge distance in polymeric dimethylberyllium (1.93 Å). The root mean square vibrational amplitude of the distances Be–C(Cp) [l = 0.075(3) Å] indicates that the Be atom is firmly bonded to the ring. The Be–C(Me), Me = CH_3, bond distance is indistinguishable from that in dimethylberyllium.

Drew and Haaland[24] have suggested that the Be atom is (sp)-hybridized and that it uses one such orbital for a two-center, two-electron bond to the methyl carbon atom. One bonding molecular orbital is then formed by combination of the other hybrid orbital with the a_1 π-orbital of the ring. Two more – degenerate – bonding molecular orbitals of higher energy are formed by combination of the two unhybridized Be $2p$-orbitals with the two e_1 π-orbitals of the ring. All these orbitals are filled and the Be atom is surrounded by an electron octet.

The molecular structure of cyclopentadienyl beryllium chloride[25] is entirely analogous to the methyl compound, but the Be–Cl bond distance [1.837(6) Å] is significantly longer than in free monomeric $BeCl_2$, where it is 1.75 ± 0.02 Å[26]. It seems reasonable to assume that the Be–Cl bonds in $BeCl_2$ are shortened through dative π-bonding, and that similar π-bonding is absent in h^5-CpBeCl since the pertinent $2p$ orbitals on Be are already involved in bonding to the ring.

Biscyclopentadienylberyllium was first studied by gas-phase electron diffraction in 1963[27]; several years later the structure was refined by least-squares calculations on the intensity data[28]. The resulting structure is shown in Fig. 2. The electron scattering pattern was found to be consistent with a model of C_{5v} symmetry in which the two Cp rings are lying parallel and staggered with a perpendicular inter-ring distance of 3.37 Å. The Be atom lies on the C_5 axis between the two rings, but not at the midpoint. The perpendicular distance from the metal atom to the nearest ring (1.47 Å) is similar to, or slightly shorter than the corresponding distance in MeBeCp, but the Be–C(1) vibrational amplitude l = 0.098(4) Å is greater than in MeBeCp, indicating that the Be atom is less firmly held in the equilibrium position. The distance to the other ring is 1.90 Å and the Be–C(6) vibrational amplitude is l = 0.115(7) Å.

This asymmetric structure explains the dipole moment of Cp_2Be in cyclohexane = 2.24 ± 0.09 Debye[29].

---

[c] The standard deviations quoted in Refs. [24] and [25] have been expanded to include an estimated uncertainty of 0.1% in the electron wavelength.

Even though the Cp rings are at different distances from the metal atom and must be assumed to be bonded to it in different ways, the difference between their structures (*e.g.* their C–C bond distances) appears to be negligible. Except for the Be atom, the molecule therefore has D_{5d} symmetry, or very nearly so. It has been suggested that the potential energy surface of Be has two minima on the C_5 axis, 0.22 Å on either side of the midpoint between the rings, and that the Be atom alternates between the two positions[27]. Indeed, ^1H NMR spectra of Cp_2Be in TMS, cyclohexane, and methylcyclohexane were later found to consist of a sharp singlet over a wide temperature range[30], showing that alternation is rapid on the NMR time scale. It has been estimated that the time spent in each minimum is 10^{-12} to 10^{-13} sec[31].

The asymmetric structure of Cp_2Be can be rationalized in the following way: optimal Be-to-Cp bonding would be achieved in a symmetrical molecule in which both perpendicular Be-to-ring distances are similar to that in CH_3BeCp, *i.e.* about 1.5 Å. This would, however, require a perpendicular ring-to-ring distance of about 3.0 Å, and such a close approach is prevented by inter-ring repulsion, since the van der Waals thickness of an aromatic ring is 3.4 Å. The two potential energy surfaces, which describe the bonding between the Be atom and each of the two rings and which add up to the total binding energy, therefore have their minima at different places, and these minima are sufficiently narrow to give two minima in the total energy surface.

In 1970 McVicker and Morgan published the infrared absorption spectra of gaseous and solid Cp_2Be[32]. The gas-phase spectrum is very simple, consisting mainly of the very few bonds expected for Cp rings of D_{5h} symmetry, while the crystal spectra are much more complex. This difference was interpreted as evidence for a change in the nature of the metal-to-ring bonding on passing from the gas phase to the crystal phase.

Very recently the crystal structure of Cp_2Be has been determined at room temperature and at –120 °C[33, 34]. The crystals are disordered at both temperatures, but the molecular structure may be described as follows: the Be atom lies about 1.5 Å above one ring on the (approximate?) fivefold symmetry axis. The second ring is parallel – or nearly parallel – to the first and lies about 1.9 Å above the Be atom, but has slipped about 1.2 Å sideways, with the result that the Be atom lies under the *edge* of the ring. The orientation of this second ring appears to change with temperature: at –120 °C the Be atom appears to lie directly beneath a C atom and at room temperature beneath the midpoint of a C–C bond. The structure is perhaps best described as a slip sandwich.

It has been suggested that the large dipole moment of Cp_2Be leads to coupling of the oscillations of the Be atoms in neighboring molecules in the crystal, which then causes the oscillations to slow down as the temperature falls. Since the bonding between the Be atom and the far ring is weak and not strongly directed, intermolecular forces would be sufficient to explain the 1.2 Å sideways slip[35].

3. Magnesium

The electron scattering pattern from biscyclopentadienyl magnesium, $(h^5\text{–}Cp)_2Mg$, is consistent with molecular models of both D_{5h} (eclipsed rings) and D_{5d} (staggered rings) symmetry, though the former gives the better fit[36]. An asymmetric structure similar to that of the Be analog can be ruled out. The Mg–C bond distance is 2.339(4) Å; this corresponds to a perpendicular ring-to-ring distance of 4.02 Å, which is considerably more than the van der Waals thickness of an aromatic ring. The Mg–C vibrational amplitude is l = 0.103(3) Å, which is very similar to the two Be–C amplitudes in Cp_2Be. The C–C distance [1.423(2) Å] is very similar to the mean C–C bond distance in Cp_2Be.

4. Boron

The molecular structures of diborane[37], tetramethyldiborane[38] and trimethylborane[39] as determined by gas-phase electron diffraction by Bartell and Carroll are shown in Fig. 3. It is interesting to note that the B–B distance in diborane is only 0.01 Å greater than twice the tetrahedral covalent radius of boron, while the B–H bridge distance is 0.14 Å longer than the terminal B–H bond: As in the case of crystalline dimethylberyllium, the internuclear distances suggest that there is bonding (though not necessarily σ-bonding) between the two metal atoms. The B–H–B bridge bond should perhaps be considered a central rather than an open three-center bond[40, 41].

Such a view is in agreement with the overlap populations obtained by *ab-initio* self-consistent-field molecular orbital calculations[42]: The overlap populations obtained for the B–B, B–H_b (b = bridge) and B–H_t (t = terminal) bonds are about 0.30, 0.40 and 0.82 respectively.

The B–B bond in tetramethyldiborane is 0.07 Å longer than in diborane and the B–C bonds are about 0.01 Å longer than in trimethylborane. These elongations are probably best explained as the result of steric repulsion[38]; the shortest distance between carbon atoms in terminal methyl groups on different boron atoms is 3.43 Å, considerably less than the van der Waals diameter of a methyl group.

It has sometimes been suggested that the reason why trimethylborane (in contrast to trimethylaluminum) is monomeric in all phases is that the monomer is stabilized through hyperconjugation. The structure of the monomer offers no support for this view. The B–C distance is in fact 0.06 Å longer than the B–C single bond distance calculated from

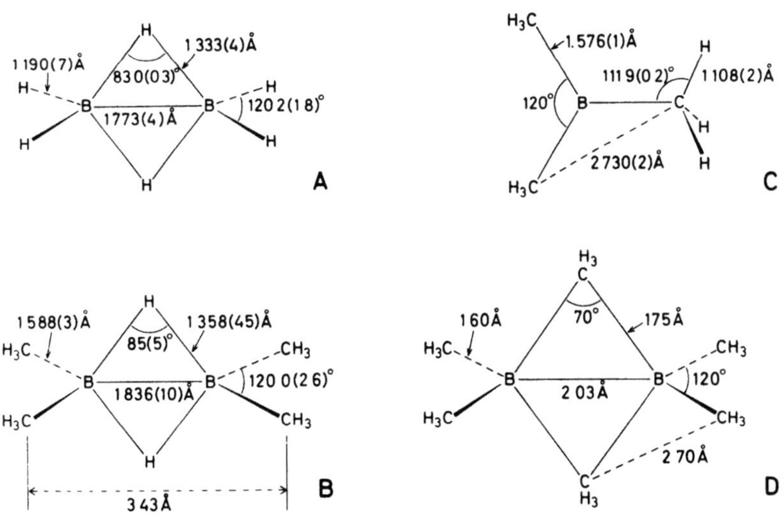

Fig. 3. The molecular structures of diborane[37] (*A*), tetramethyldiborane[38] (*B*) and trimethylborane[39] (*C*) in the gas phase and of the hypothetical trimethylborane dimer (*D*)

Pauling's revised Schomaker-Stevenson rule[39]. Rather, it is clear that steric repulsion alone is sufficient to explain why no methyl-bridged dimer forms: the boron atom is too small to accommodate four methyl groups. In a hypothetical dimer (Fig. 3d) the eight $C_b \cdots C_t$ distances would be only about 2.70 Å. In dimeric trimethylaluminum the corresponding distance is 3.23 Å.[d)]

For trimethylborane, theoretical RD functions were calculated for distributions of methyl group orientations with the use of the classical probability distribution $P(\phi) = A \exp[-V(\phi)/kT]$, where ϕ is the angle of rotation of the methyl group and $V(\phi) = \frac{1}{2} V_0 [1 - \cos(6\phi)]$, since a sixfold barrier is the simplest form compatible with the molecular geometry. The best fit between experimental and theoretical curves was obtained with $V_0 = 0$, although barriers of several hundred calories could not be ruled out.

For tetramethyldiborane, the equilibrium orientation of the methyl groups appears to be the one in which the C–H bonds are staggered with respect to the B–C and the two $B{-}H_b$ bonds radiating from the boron atom. A potential $V = \frac{1}{2} V_0 [1 - \cos(3\phi)]$ was assumed, and the best fit was obtained for $V_0 = 1$ kcal mole^{-1}.

5. Aluminum[e)]

Gaseous trimethylaluminum consists of monomeric and dimeric species in a temperature- and pressure-dependent equilibrium. Thus, gas at 215 °C and 30 mm consists of more than 96% monomers, while gas at 60 °C and the same pressure consists of more than 97%

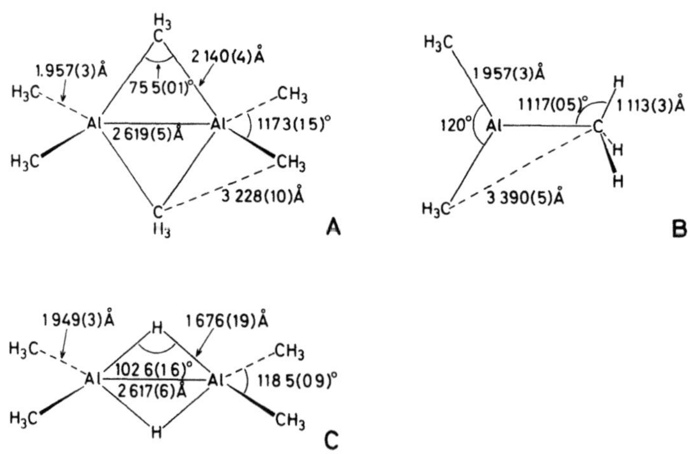

Fig. 4. The molecular structures of dimeric[48)] (*A*) and monomeric[48)] (*B*) trimethylaluminum and dimeric dimethylaluminum hydride[57)] (*C*) in the gas phase

---

[d)] Similarly, the small size of the boron atom may account for BH_3 forming a volatile dimer, while AlH_3 forms a nonvolatile polymer in which each aluminum atom is surrounded by six bridging hydrogen atoms[43)].

[e)] For recent reviews of organoaluminum chemistry see Refs. [9)] and [44)].

dimers. The enthalpy of dissociation of one mole of gaseous dimer into two moles of gaseous monomer is $\Delta H_{d(l)}$ = 20.40 ± 0.34 kcal/mole[45]. The concentration of monomers in hydrocarbon solutions at room temperature is negligible, and crystalline trimethylaluminum consists of dimeric species only. The diborane-like, methyl-bridged structure of the dimer (see Fig. 4) was established by an X-ray diffraction study by Lewis and Rundle[46]. In a low-temperature crystal study Huffman and Streib[47] recently succeeded in locating the hydrogen atoms: the three C–H bonds in a bridging methyl group and the vector $C_b \cdots C_b$ point towards the four corners of slightly distorted tetrahedron. Hence, it is clear that the hydrogen atoms are not involved in bonding to the aluminum atoms.

The bond distances and valence angles of dimeric trimethylaluminum given in Fig. 4 are those obtained by gas-phase electron diffraction[48] at a nozzle temperature of about 60 °C. The bond distances are in excellent agreement with those obtained in the low-temperature crystal study: Al–C_t = 1.949(2) Å and 1.956(2) Å, Al–C_b = 2.123(2) Å and 2.152(2) Å, and Al–Al = 2.606(2) Å, particularly since correction for thermal motion in the crystal would add a few thousandths of an Å to the latter. The < Al–C_b–Al angle 75.7(0.1)° found in the crystal is also very similar to the ED result. The < C_t–Al–C_t angle, however, = 123.2(0.1)°, differs significantly from that found in the gas phase. The difference may be due to crystal packing forces.

In this compound, too, the internuclear distances indicate that bonding between the metal atoms is important: The Al–Al distance is only 0.10 Å more than the value calculated for a single bond by doubling the tetrahedral covalent radius, and 0.24 Å less than the Al–Al distance in the metal[49]. The Al–C_b bonds on the other hand are 0.18 Å longer than the Al–C_t bonds. The latter are equal to the Al–C single-bond value 1.96 Å, calculated from the tetrahedral covalent radii of carbon and aluminum and the revised Schomaker-Stevenson rule.

Approximate self-consistent-field molecular orbital calculations do indeed indicate considerable Al–Al bonding of the $3p\sigma$–$3p\sigma$, $3p\pi$–$3p\pi$ type[50].

The ^{1}H NMR spectra of trimethylaluminum dimer in hydrocarbon solutions at room temperature consist of one peak only, showing that exchange of terminal and bridging methyl groups is rapid on the NMR time scale[51]. There is now general agreement that the exchange is effected through dissociation into monomeric species followed by fast recombination[44, 52].

The molecular structure of trimethylaluminum monomer[48] has been determined by gas-phase electron diffraction at a nozzle temperature of about 215 °C. The structure is entirely analogous to that of trimethylborane, the molecular symmetry, except for hydrogen atoms, is D_{3h} and the barrier to rotation of the methyl groups is negligible. The Al–C bond distance is equal to the terminal Al–C distance in the dimer, and no shortening that might be ascribed to hyperconjugation is detected.

The methyl groups in trimethylaluminum dimer are closely packed, the eight $C_b \cdots C_t$ distances being 3.23 Å and, as in the case of dialkylberyllium compounds, the tendency to associate decreases when the methyl groups are replaced by more bulky substituents. Thus the heat of dissociation of the liquid dimers decreases from 19.40 ± 0.30 kcal mole^{-1} for trimethylaluminum[54] to 16.93 ± 0.23 kcal mole^{-1} for triethylaluminum[55] to 8.16 ± 0.12 kcal mole^{-1} for tri*iso*butylaluminum[56]. The latter is only 40% associated as a pure liquid at 10°.

Dimethylaluminum hydride is trimeric in hydrocarbon solution at room temperature[56]. The gas is dimeric (by molecular weight) at 167 °C and 55 mm; at 83 °C and 56 mm the average degree of association is 2.5[56]. The molecular structure of the dimer (Fig. 4) has been determined by gas-phase electron diffraction at a nozzle temperature of about 170 °C[57]. Attempts to determine the molecular structure of the trimer by recording the electron-scattering pattern at a nozzle temperature of 65 °C were not successful; the scattering pattern showed that the concentration of trimer in the gas jet must have been negligible[58].

The structure of dimeric dimethylaluminum hydride is remarkably similar to that of trimethylaluminum dimer. No significant differences are found between either the Al—Al and Al—C_t bond distances or the $< C_t$—Al—C_t angles in the two compounds. The constancy of the Al—Al distance is most easily rationalized if direct bonding is assumed. The Al—H_b distance is considerably longer than the (terminal) Al—H bond in $H_3AlN(CH_3)_3$ = 1.560(11) Å[59].

The molecular structure of trimeric dimethylaluminum hydride can only be a matter for speculation. If metal-metal bonding is important, the Al_3H_3 ring must be planar. This is in agreement with the infrared absorption spectrum of dimethylaluminum hydride in cyclohexane[60]. If the Al—H_b and Al—Al bond distances are equal to those in the dimer, the $< H_b$—Al—H_b angle must be 137°.

Dimethyl(cyclopentadienyl)aluminum[61] is a colorless solid at room temperature. The melting point is high (about 140 °C), the vapor pressure low (about 0.001 mm at 60 °C), and the solubility in hydrocarbon solvents very poor[62]. The degree of association in freezing benzene is 1.4. The electron scattering pattern from the gas, recorded at a nozzle temperature of 130 °C, showed that the jet contained monomeric species only[63]. It was found that molecular models of the monomer containing σ-bonded or symmetrically π-bonded cyclopentadienyl rings could be ruled out, since both models lead to serious disagreement between experimental and calculated radial distribution curves in the region from 1.9 to 2.3 Å, *i.e.* in the region of Al—C bond distances. The four models containing asymmetrically π-bonded rings shown in Fig. 5 were then considered, and it was found that all of them could be brought into satisfactory agreement with the electron-diffraction data. In each case the value obtained for the distance between the Al atom and the (approximate) fivefold symmetry axis of the ring (o—p in Fig. 5) was such that the Al atom is situated directly above the edge of the ring, above the midpoint of the C_1—C_2 bond in models I and III and above C_1 in models II and IV. The perpendicular distances h = Al—p obtained ranged from 2.06 to 2.19 Å.

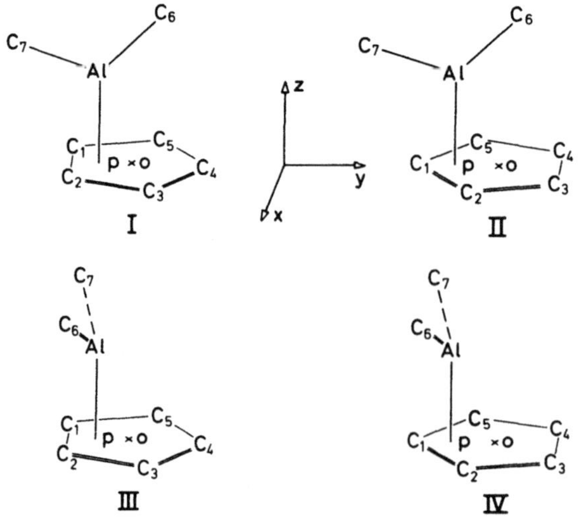

Fig. 5. Molecular models for dimethyl(cyclopentadienyl)aluminum in the gas phase[63]

Subsequent approximate self-consistent-field molecular orbital calculations[63] indicated that the most stable conformer is I, but that the barrier to rotation of the cyclopentadienyl ring (*via* II) is of the order of 5 kcal $mole^{-1}$ or less and that the barrier to exchange of the methyl groups (*via* III) is between 10 and 20 kcal $mole^{-1}$.

The most important structure parameters obtained by refinement on model I were: Al–p = 2.10(2) Å, o–p = 0.99(10) Å, Al–C_1 = 2.21(2) Å, Al–C_6 = 1.952(3) Å, C–C (mean) = 1.422(2) Å and $< C_6$–Al–C_7 = 124(3)°. Thus the Al–C_6 bond distance and the $< C_6$–Al–C_7 angle are not significantly different from the Al–C_t distance and $< C_t$–Al–C_t angle in dimeric trimethylaluminum, and the Al–C_1 distance is only slightly longer than the Al–C_b distance.

It would seem that dimethyl(cyclopentadienyl)aluminum in the gas phase is best described as containing *di-hapto* rings: $(CH_3)_2Al(h^2\text{–}C_5H_5)$.

The ^1H NMR spectrum of $(CH_3)_2Al(C_5H_5)$ in benzene at ambient temperature consists of one singlet corresponding to the six methyl protons and another singlet corresponding to the five cyclopentadienyl protons[61]. This is, of course, in agreement with the low barrier to rotation of the ring and the moderate barrier to exchange of the methyl groups obtained

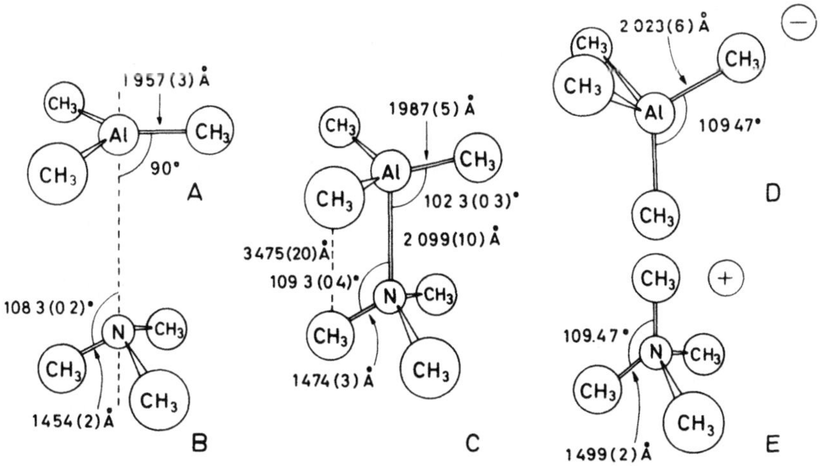

Fig. 6. The molecular structures of monomeric trimethylaluminum[48] (*A*), trimethylamine[65] (*B*) and trimethylaluminum-trimethylamine[65] (*C*) in the gas phase, and idealized structures of the tetramethylaluminate ion (*D*) (the Al–C bond distance has been taken from the crystal structure of $Li[Al(C_2H_5)_4]$[67]) and the tetramethylammonium ion (*E*) (the N–C bond distance has been taken from the crystal structure of $[(CH_3)_4N]F\text{–}4H_2O$[68])

by the molecular orbital calculations. It should be kept in mind, however, that the compound is partly associated in benzene, and therefore there may be other mechanisms of exchange.

Haaland and Weidlein have recorded the infrared and Raman spectra of solid $(CH_3)_2Al(C_5H_5)$[62]. The spectra provided strong evidence against the presence of bridging methyl groups and σ-bonded cyclopentadienyl groups. They therefore proposed a polymeric structure containing bridging cyclopentadienyl groups of approximate D_{5h} symmetry. The proposal was later somewhat modified[63] to provide for the sort of bridging cyclopentadienyl rings found in solid $(C_5H_5)_3In$[64]. This compound is polymeric in the solid phase. Each indium atom is surrounded by two terminal σ-bonded cyclopentadienyl rings and two bridging rings of approximate D_{5h} symmetry. Each bridging ring is bonded to two In atoms lying on opposite sides of the ring plane. Only two contacts, In–C_1 and In'–C_3, are short enough to indicate appreciable bonding.

Trimethylaluminum forms stable complexes with a number of Lewis bases like trimethylamine, trimethylphosphine and dimethyl ether. The molecular structure of trimethylaluminum-trimethylamine as determined by gas-phase electron diffraction is shown in Fig. 6[65]. Also shown in Fig. 6 are the molecular structures of the free acceptor[48] and the free donor[66] (both determined by gas-phase electron diffraction) and the idealized structures of the tetramethylaluminate and tetramethylammonium ions. The Al–C bond distance in the former has been taken from the crystal structure of $Li|Al(C_2H_5)_4|$[67], and the N–C bond distance in the latter from the crystal structure of $|(CH_3)_4N|F \cdot 4\,H_2O$[68].

It is seen that the structure of the acceptor in the complex is intermediate between that of the free acceptor and the corresponding "-ate" ion (which may be regarded as a complex with the very strong donor CH_3^-). The Al–C bond distance increases from 1.96 Å in the free acceptor to 1.99 Å in the complex and 2.02 Å in the ion. The angle between the threefold symmetry axes and the Al–C bonds increases from 90° in the planar free acceptor to 102° in the complex and 109° in the "-ate" ion.

This variation can be rationalized in terms of the valence-shell electron-pair repulsion theory[8]: introduction of electrons into the formerly empty $3p\pi$ orbital of the free acceptor leads to repulsion of the electron pairs of the three Al–C bonds.

Similarly, it might be argued that since complex formation is accompanied by transfer of electrons away from the donor, the bonding and the structure of the donor in the complex should be intermediate between that of the isolated donor and the corresponding positive ion $|(CH_3)_4N|^+$, which may be regarded as a complex of trimethylamine with CH_3^+. Inspection of Fig. 6 shows that this is indeed the case: the N–C bond distance increases from 1.45 Å in the free donor to 1.47 Å in the complex and 1.50 Å in the ion. The angle between the threefold symmetry axis and the N–C bonds is slightly less than tetrahedral in the free donor and increases somewhat on formation of the complex, though the difference is of marginal statistical significance.

The observed variations in the N–C bond distances and $< C_3$–N–C angles are *not* in agreement with the valence-shell electron-pair repulsion model. This predicts that partial removal of the lone-pair electrons on the nitrogen atom would lead to a shortening of the N–C bonds. It would further predict that $< C_3$–N–C should be greater than tetrahedral in the free donor and decrease on complex formation.

The variation can, however, be understood if it is assumed that repulsive interactions between the atoms or groups bonded to the nitrogen atom predominate over the valence-shell electron-pair repulsion. That repulsion between the substituents should dominate does not seem unreasonable, since the nitrogen atoms in these species are bonded to three or four atoms, all with a larger covalent radius than the nitrogen.

The molecular structure of the complex trimethylamine-borane[69] has been determined by microwave spectroscopy. Here too, the N–C bond distance (1.495 Å) was found to be significantly longer than in free trimethylamine, and the $< C_3$–N–C angle (110.9°) nearly tetrahedral.

The P–C bond distance in trimethylphosphine-borane[70] on the other hand is *shorter* than in free trimethylphosphine[71]; the distances are P–C = 1.819 ± 0.010 Å and 1.843 ± 0.003 Å, respectively, as determined by microwave spectroscopy. We are not aware of any accurate determination of the structure of a tetraalkylphosphonium ion, but in the $[(C_6H_5)_3PCH_3]^+$ ion[72] the P–C(Me) distance is 1.779(3) Å by X-ray diffraction. The C_3–P–C angle *decreases* from 118.7 ± 0.02° in the free donor to 113.6 ± 0.4° in $H_3BP(CH_3)_3$.

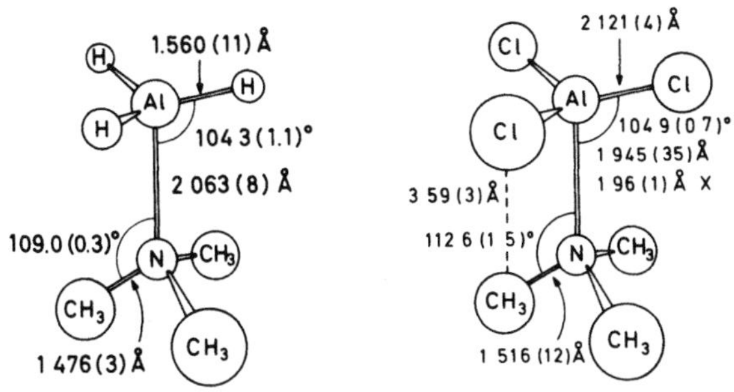

Fig. 7. The molecular structures of trimethylamine-alane[59] and trichloro(trimethylamine)aluminum[73] in the gas phase

In these species the substituent atoms are *smaller* than the phosphorus atom and the variation in bond distances and angles is as predicted by the valence-shell electron-pair repulsion theory.

The molecular structures of trimethylamine-alane[59] and trichlorotrimethylamine-aluminum[73], as determined by gas-phase electron diffraction, are shown in Fig. 7. The latter compound has also been studied by X-ray diffraction[74], and the Al–N bond distance obtained in that study [1.96(1) Å] is included in the Fig. 7. The other parameters were more accurately determined by electron diffraction.

It is seen that the Al–N distance, which in $(CH_3)_3AlN(CH_3)_3$ was 2.10 Å, decreases to 2.06 Å in $H_3AlN(CH_3)_3$ and to 1.96 Å in $Cl_3AlN(CH_3)_3$. The variation is most easily explained as an inductive effect; the introduction of more electronegative substituents of Al increases the positive charge on the metal atom and enhances complex formation. Indeed, approximate self-consistent-field molecular orbital calculations[75] indicates that the charge on the aluminum atom is positive in all three complexes, that the net negative charge on the acceptor is consequently carried by the substituents, and that the charge transferred from donor to acceptor increases in the sequence

$$(CH_3)_3AlN(CH_3)_3 < H_3AlN(CH_3)_3 < Cl_3AlN(CH_3)_3.$$

In agreement with this, the N–C bond distance in $Cl_3AlN(CH_3)_3$ is found to be significantly longer than in the two other complexes and is indistinguishable from the N–C bond distance in the $[(CH_3)_4N]^+$ ion.

The Al–Cl bond distance in $Cl_3AlN(CH_3)_3$ is significantly longer than in free monomeric aluminum trichloride[76], 2.06(1) Å by gas-phase electron diffraction, and longer than the mean Al–Cl bond distance in the complex of aluminum trichloride with propionyl chloride[77], which is 2.093(3) Å. (Propionyl chloride is a weaker donor than trimethyl-amine.)

Substituted dialkylalanes of the type R_2AlX, where X is an atom or a group with lone-pair electrons like NR'_2, OR′ or F, associate through the formation of N, O, or F bridges. The degree of association appears to be determined primarily by the nature of the bridging atom. Thus dimethylaluminum fluoride[78] and diethylaluminum fluoride[79] are both tetrameric, dimethylaluminum methoxide[80] and diethylaluminum methoxide[81] are both trimeric, and dimethylaluminum methylamide[82], $(CH_3)_2AlNHCH_3$, forms trimers with *cis* or *trans* arrangement of the hydrogen atoms. The degrees of association, *i.e.* $n = 3$ for nitrogen and oxygen bridges and $n = 4$ for fluorine bridges, may be regarded as "normal".

Table 1. Structure parameters of associated organoaluminum compounds of the type $[(CH_3)_2AlX]_n$

X	n	Al–C(Å)	Al–X(Å)	Al···Al(Å)	<Al–X–Al (deg)	<X–Al–X (deg)	<C–Al–C (deg)	Ref.
F	4	1.947(4)	1.810(3)	3.463(10)	146.1(2.6)	92.3(1.2)	131.2(1.9)	86)
OCH_3	3	1.957(3)	1.851(3)	3.297(7)	125.8(0.4)	103.2(1.1)	117.3(0.8)	85)
$NHCH_3$	3	1.973(6)	1.940(5)	3.399(7)	122.3(0.4)	102.1(0.4)	116.9(0.4)	82)
$OC(CH_3)$	2	1.962(7)	1.864(3)	2.816(8)	98.1(0.3)	81.9(0.3)	121.7(0.9)	89)
$N(CH_3)_2$	2	1.951(8)	1.958(5)	2.809(4)	91.7(0.2)	88.3(0.3)	115.7(0.5)	82)

But if the alkyl groups, R or R′, are sufficiently bulky, steric interaction between them may limit the degree of association actually obtained to values less than "normal". Thus di-*iso*-propylaluminum fluoride[83] is trimeric, dimethylaluminum phenoxide equilibrates to a mixture of dimers and trimers[81], dimethylaluminum-*tert*-butoxide is dimeric[84], and so is dimethylaluminum dimethylamide[80].

The molecular structures of *cis* and *trans* $[(CH_3)_2AlNH(CH_3)]_3$ have been determined by X-ray diffraction[82]. Both isomers contain nonplanar Al_3N_3 rings. The *cis*-isomer crystallizes in a chair conformation of approximate C_{3v} symmetry. The three N–H bonds are all axial. The most important bond distances and valence angles are listed in Table 1. The *trans* isomer, on the other hand, crystallizes in a skew-boat conformation of C_2 symmetry. This change is easily explained as the effect of the considerable repulsion that would occur with methyl groups bonded axially to nitrogen if the molecules were in a chair conformation. The distance between the nitrogen atoms is only 3.02 Å.

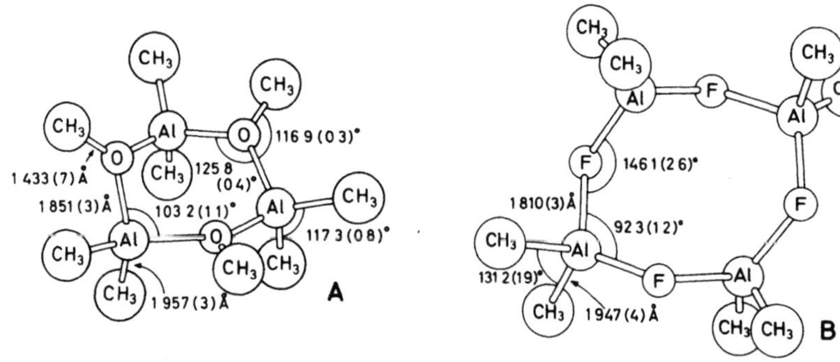

Fig. 8. The molecular structures of trimeric dimethylaluminum methoxide[85] (*A*) and tetrameric dimethylaluminum fluoride[86] (*B*) in the gas phase

Trimeric dimethylaluminum methoxide[85] and tetrameric dimethylaluminum fluoride[86] have been studied by gas-phase electron diffraction (see Fig. 8). The most important bond distances and valence angles are also listed in Table 1. Both compounds contain non-planar rings. The electron scattering pattern from $[(CH_3)_2AlOCH_3]_3$ was consistent with a chair model of C_{3v} symmetry, but models of lower symmetry were not ruled out. The three valencies of the oxygen atoms are lying in a plane. When the molecule is in the chair conformation, the six distances from methyl groups bonded to oxygen to methyl groups bonded equatorially to aluminum is only 3.20 Å. The strain would be somewhat reduced if the molecule changed its conformation, but there will always be six methyl(0) to methyl(Al) distances that are 3.50 Å or less.

The electron-scattering pattern from $[(CH_3)_2AlF]_4$ is inconsistent with molecular models of C_{4v} or D_{2d} symmetry. A chair-boat model of C_s symmetry was consistent with the data, but other models of low symmetry were not ruled out. In this model there is little van der Waals contact between methyl groups bonded to different aluminum atoms.

The endocyclic valence angels of $[(CH_3)_2AlNHCH_3]_3$, $[(CH_3)_2AlOCH_3]$, and $[(CH_3)_2AlF]_4$ suggest that the "normal" degree of association achieved in the absence of excessive strain between substituents is determined by the need to form rings that are relatively free from angle and Pitzer strain. The formation of dimers in which the average endocyclic angle must be 90° or less is only possible if accompanied by large deformations of the < Al–N–Al, < Al–O–Al, or < Al–F–Al angles. The magnitude of the endocyclic angles of $[(CH_3)_2AlF]_4$ shows that it would be possible to form a trimer free from angle strain. Such a trimer would, however, have to be very nearly planar and the Al–F bonds around the ring would be almost eclipsed. It seems reasonable to assume that the greater stability of the tetramer is due to the existence of a significant barrier to rotation about the Al–F bonds and the severe Pitzer strain that this would introduce into the near-planar trimer.

The deviation from tetrahedral valence angles at the aluminum atoms in the three species under discussion is adequately explained by the valence-shell electron-pair repulsion theory. The very polar Al–X bonds should require less space than the more covalent Al–C bonds, and the difference should be most pronounced in the fluoride.

The wide < Al–N–Al angle in $[(CH_3)_2AlNHCH_3]_3$ can also be rationalized in terms of the valence-shell electron-pair repulsion theory, but the increase of the endocyclic angle on the bridging atom on going to $[(CH_3)_2AlOCH_3]_3$ and $[(CH_3)_2AlF]_4$ cannot. The theory predicts that the angle should decrease and be less than tetrahedral in the fluoride. The situation is analogous to that encountered in the discussion of the structure of $(CH_3)_3AlN(CH_3)_3$, where the geometry around the acceptor atom could be adequately rationalized in terms of the valence-shell electron-pair repulsion theory, while the geometry around the donor could not.

The wide < Al–O–Al and < Al–F–Al angles as well as the planarity of the oxygen atom might be rationalized by evoking dative $p\pi$–$d\pi$ bonding, as has been done to rationalize the wide < Si–O–Si and < Si–N–Si angles encountered when an oxygen or nitrogen atom is bonded between two silicon atoms[f]. It is likely to be some time before the importance of the 3*d* atomic orbitals in determining the structure and reactions of organoaluminum compounds can be settled by accurate molecular orbital calculations. In the meantime we prefer to look for other ways to rationalize the wide < Al–O–Al and < Al–F–Al angles. We can, for instance, assume that repulsion between the substituents predominates over repulsion between the valence-shell electrons when two or more large atoms (like Al) are bonded to a small atom (like N, O or F).

If < Al–N–Al, < Al–O–Al and < Al–F–Al are all assumed to be tetrahedral, the resulting Al···Al distances would be 3.17, 3.02 and 2.95 Å, respectively, which is only slightly longer than in metallic aluminum. The observed Al···Al distances are 3.40, 3.30 and 3.46 Å, respectively. While the lowest-energy three-center orbital of an Al–X–Al bridge is bonding between the aluminum atoms, the second lowest is antibonding[88]. In electron-deficient compounds like the trimethylaluminum dimer or dimeric dimethyl-

---

[f] For a critical review see Ref.[87].

aluminum hydride only the lowest orbital is occupied. But in electron-precise compounds like

$$[(CH_3)_2AlNHCH_3]_3,\ [(CH_3)_2AlOCH_3]_3 \text{ and } [(CH_3)_2AlF]_4$$

both orbitals are occupied, and the result is probably a net antibonding effect. When the metal atoms are much larger than the bridging atom, the antibonding effect would be particularly large and so would repulsion between the metal atom cores. In such a case an unusually wide < Al–X–Al angle might be expected.

Similarly, the planarity of the oxygen atom in $[(CH_3)_2AlOCH_3]_3$ may be explained as the result of repulsion between the methyl group and the two aluminum atoms, the Al · · · C(0) distance is only 2.81 Å.

Two compounds of the type R_2AlX, where the degree of association is less than "normal", have been studied: $[(CH_3)_2AlN(CH_3)_2]_2$ by X-ray diffraction[82] and $[(CH_3)_2AlOC(CH_3)_3]_2$ by electron diffraction[89]. The compounds contain planar Al_2N_2 and Al_2O_2 rings and the valences of the oxygen atoms are lying in a plane. The most important bond distances and valence angles are listed in Table 1.

It is evident that both species are highly strained: The endocyclic angles at the bridging atoms are 30° less than in the strainfree trimers. The Al · · · Al distance is 2.81 Å in both compounds, *i.e.* 0.04 Å *less* than in metallic aluminum. In both compounds the angle at the bridging atom is greater than 90°; it is greatest in $[(CH_3)_2AlOC(CH_3)_3]_2$ where the Al–X bond distance is shortest. The endocyclic angles at the aluminum atom are consequently smaller than 90°, but the exocyclic < C–Al–C angles differ little from the angles in the unstrained trimers. It seems reasonable to assume that the Al–X bonds are highly bent.

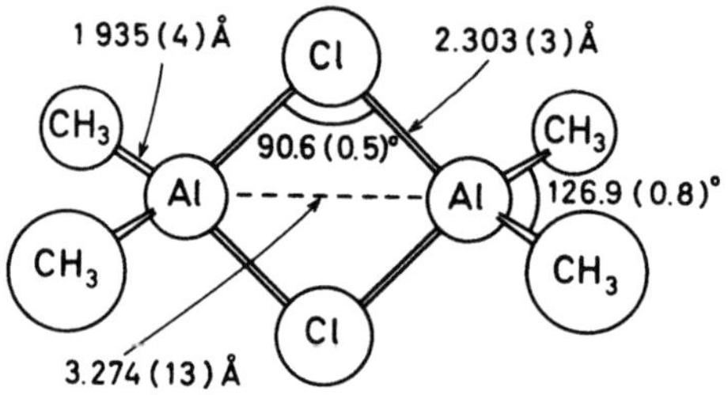

Fig. 9. The molecular structure of dimeric dimethylaluminum chloride in the gas phase[90]

The molecular structure of the dimethylaluminum chloride dimer as determined by gas-phase electron diffraction[90] is shown in Fig. 9. Since the molecule contains a planar four-membered ring, it cannot be assumed to be free of angle strain. However, the fact that the compound is dimeric rather than trimeric or tetrameric allows us to conclude that the angle strain is modest and that the endocyclic angles differ little from their preferred value.

While the valence angles around the aluminum atom are indistinguishable from those in $[(CH_3)_2AlF]_4$, the < Al–Cl–Al angle is only 91°. That this angle is less than tetrahedral is in agreement with the valence-shell electron-pair repulsion model. Clearly the greater size of the chlorine atom reduces the importance of repulsion between the metal atoms.

The Al···Al distance is 0.19 Å less than in the fluoride; an < Al–Cl–Al angle of 100° would make it 0.07 Å longer.

Dimethylaluminum thiomethoxide, $(CH_3)_2AlS(CH_3)$, is polymeric in the crystal and dimeric in the gas phase. In the crystal it forms infinite zigzag chains of alternating aluminum and sulfur atoms[91], < Al–S–Al = 103.0(0.1)° and < S–Al–S = 100.1(0.1)°. The three valences of the sulfur atom are *not* lying in a plane and the < C–S–Al angles are less than tetrahedral. The Al–S bond distance is 2.345(2) Å.

The dimer is presently being studied by gas-phase electron diffraction[92]. The Al_2S_2 ring is planar and < Al–S–Al is 95°. That the endocyclic angle at the bridging atom is greater than in dimeric dimethylaluminum chloride is in accordance with the valence-shell electron-pair repulsion theory. On going to dimethylaluminum dimethylphosphide, $(CH_3)_2AlP(CH_3)_2$, one would expect a further opening of the angle at the bridging atom. This may be the reason why this compound forms trimers[80] rather than dimers.

6. Cyclopentadienyl Derivatives of other Main-Group Elements

Cyclopentadienylindium, $(h_5\text{–}C_5H_5)In$, has been investigated by gas-phase electron diffraction[93] and found to have the half-sandwich structure with C_{5v} symmetry sketched in Fig. 10 *A*. One may assume the bonding in this compound to be analogous to that in $(C_5H_5)BeCH_3$, that is, one may assume the In atom to be (*sp*)-hybridized, a lone pair of

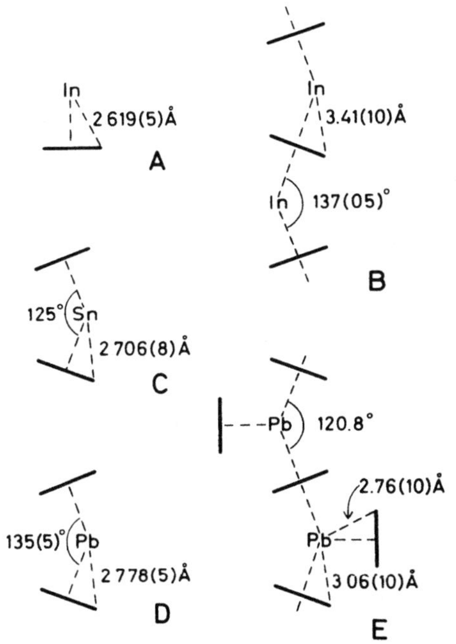

Fig. 10. The molecular structures of cyclopentadienylindium[93] (*A*), biscyclopentadienyltin[95] (*C*) and biscyclopentadienyllead[95] (*D*) in the gas phase and of cyclopentadienylindium[94] (*B*) and biscyclopentadienyllead[96] (*E*) in the crystalline phase

electrons to be contained in one (*sp*)-hybrid, and the bonding to the cyclopentadienyl ring to be effected by the other (*sp*) hybrid and the two unhybridized 5*p* orbitals.

In the solid phase $(C_5H_5)In$ is polymeric, forming zigzag chains of alternating metal atoms and cyclopentadienyl rings[94]. (Fig. 10 *B*). The In–C bond distance (calculated from the perpendicular metal-to-ring distance given in Ref. [94] and a C–C bond distance of 1.425 Å) is 3.41(10) Å, that is, 0.8 Å longer than in the gaseous monomer. The angle between the lines connecting the indium atom with the centers of the two nearest rings is 137.0(0.5)°. It seems reasonable to assume that the indium atom is approximately (sp^2)-hybridized, that two such hybrids point towards the centers of the two nearest rings, and that the third contains an electron lone pair. The two (sp^2)-type orbitals on neighboring indium atoms pointing towards the same ring combine with the a_1 π orbital of the ring to form one three-center orbital which is bonding between the metal atoms and the ring, one which is nonbonding, and one which is antibonding. The bonding orbital is occupied by two electrons.

The structures of two bis-cyclopentadienyl derivatives of group-IV metals, $(h_5\text{–}C_5H_5)_2Sn$ and $(h_5\text{–}C_5H_5)_2Pb$, have been determined by gas-phase electron diffraction[95] (see Fig. 10 *C* and *D*). In both compounds the cyclopentadienyl rings have D_{5h} symmetry, or very nearly so, and the metal atoms are lying on or very close to the fivefolg symmetry axis of each ring. The angle between the fivefold axes is about 125° in $(C_5H_5)_2Sn$ and 135 (5)° in $(C_5H_5)_2Pb$, *i.e.* very similar to the corresponding angle in polymeric $(C_5H_5)In$. Again the metal atoms may be assumed to be roughly (sp^2)-hybridized, one hybrid containing a lone pair, and each of the other two bonding to one ring through interaction with the a_1 π orbital. The unhybridized *p* orbital may be regarded as empty, though there undoubtedly is some interaction with the appropriate occupied e_1 π orbitals of the rings.

$(C_5H_5)_2Pb$ is also polymeric in the solid phase. The structure as determined by X-ray diffraction[96] is sketched in Fig. 10 *E*. Each lead atom is surrounded by one terminal and two bridging cyclopentadienyl rings. The lines connecting the lead atom with the centers of the three rings lie in a plane. The distance from the metal atom to the carbon atoms in the terminal ring is not significantly different from the corresponding distance in the gaseous monomer, but the distance to the carbon atoms in the bridging cyclopentadienyl groups is about 0.3 Å more. It was suggested[96] that one (sp^2)-type hybrid effects bonding to the terminal cyclopentadienyl ring through interaction with the a_1 π orbital, as in the monomer, that the a_1 π-orbital on each ring combines with the appropriate hybrids on the two neighboring metal atoms to produce three center orbitals in the same manner as proposed for polymeric cyclopentadienylindium, and that the two lowest orbitals are occupied. The unhybridized 6*p* orbital may again be regarded as empty, though there undoubtedly is interaction with the appropriate e_1 π orbitals of the neighboring rings.

Electron diffraction investigations of trimethyl silylcyclopentadiene[97], $(CH_3)_3Si(C_5H_5)$, and the analogous germanium[98] and tin[99] compounds have shown that these compounds contain σ-bonded cyclopentadienyl rings. Because of the low symmetry of these compounds only a limited number of parameters could be refined; the remainder were fixed at assumed values. Nevertheless, it was concluded that the cyclopentadienyl rings in these compounds were nonplanar, in contrast to cyclopentadiene itself, which is planar[100].

However, in a recent gas-phase electron-diffraction investigation of silylcyclopentadiene[101] the ring was found to be planar. This compound is considerably simpler than trimethylsilylcyclopentadiene, consequently more parameters could be refined and the result must be considered more reliable. In our view, more work must be done on $(CH_3)_3Si(C_5H_5)$ and the germanium and tin analogues before the nonplanarity of the rings in these substances can be considered established.

7. Ferrocene and other Biscyclopentadienyl Derivatives of the Transition Elements

The molecular structure of ferrocene, $(C_5H_5)_2Fe$, as determined by gas-phase electron diffraction[102, 103] is shown in Fig. 11, and the main structure parameters are listed in Table 2[g] along with the corresponding parameters of biscyclopentadienyl derivatives of other transition elements and some of the cyclopentadienyl derivatives of main-group elements already mentioned. The angle between the C–H bonds and the C_5 ring plane,

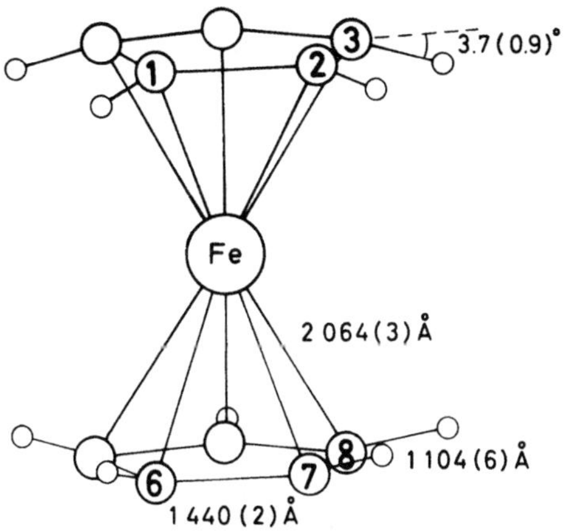

Fig. 11. The molecular structure of ferrocene in the gas phase[36, 102, 103]

Table 2. Structure parameters of biscyclopentadienylmetal compounds, $(C_5H_5)_2M$

M	M–C(Å)	*l*(M–C)(Å)	C–C(Å)	<C_5, CH(deg)	Ref.
Fe	2.064(3)	0.062(1)	1.440(2)	3.7(0.9)[1]	36, 102, 103)
Ru	2.196(3)	0.060(1)	1.439(2)	4.7(0.9)[1, 2]	103)
Cr	2.169(4)	0.078(2)	1.431(2)	2.9(1.1)[1]	36)
Ni	2.196(4)	0.084(3)	1.430(2)	0.3(1.5)[3]	113)
Mn	2.383(3)	0.135(2)	1.429(2)	6.5(1.4)[3]	114)
Pb	2.778(5)	0.142(4)	1.430(2)	0[4]	95)
Sn	2.706(8)	0.156(9)	1.431(3)	0[4]	95)
Mg	2.339(4)	0.103(3)	1.423(2)	−1.0(1.6)[1]	36)
Be	(1.907(5)	0.098(4)	1.425(2)	0[4]	28)
	2.256(7))	0.115(7)			

[1]) Corrected for shrinkage.
[2]) Shrinkage correction assumed equal to that of ferrocene.
[3]) Not corrected for shrinkage.
[4]) Assumed value.

[g]) The values listed have been taken from Ref. [103]), since the parameters quoted in Ref. [102]) are flawed by a slight scale error in part of the intensity material[103]).

$< C_5$, C–H, has been defined as positive when the C–H bonds are bent *toward* the metal atom. In the case of ferrocene it has been corrected for shrinkage[36, 104]. Even though the deformation is small, the angle is significantly greater than zero. The deformation may be rationalized by assuming that the carbon atoms are hybridized in such a way that the atomic orbital responsible for bonding to the metal atom is not a pure $2p$ orbital, but a hybrid orbital (of predominant p character) pointing its minor lobe more in the direction of the metal atom.

That the metal-to-ligand bond is particularly strong in ferrocene is shown by the root-mean-square vibrational amplitude of the M–C bond, l(Fe–C) = 0.062(1) Å, which is considerably less than in for instance $CH_3Be(C_5H_5)$.

A number of approximate molecular orbital calculations on ferrocene were reviewed by Cotton and Wilkinson in 1959[105], and the result of an *ab-initio* calculation was published in 1972[106]. The latter work also contains references to approximate molecular orbital calculations published between 1959 and 1972. It seems that the particular stability of ferrocene is due to its closed-shell structure: all occupied molecular orbitals are bonding between the metal and the ligands, the cyclopentadienyl rings function as five-electron ligands, and the iron atom attains an inert gas configuration of 18 electrons in the valence shell.

The equilibrium conformation of the free ferrocene molecule is that in which the cyclopentadienyl rings are *eclipsed* and the molecular symmetry is D_{5h}, but the barrier to internal rotation of the rings through a staggered conformation of D_{5d} symmetry is only 0.9(0.3) kcal mole^{-1}. This means that under the conditions of the electron diffraction experiment, where the nozzle temperature was 140 °C, the gas jet contained a significant number of molecules in any instantaneous conformation, eclipsed, staggered or intermediate, the number of staggered molecules being only one third the number of eclipsed[103].

Satisfactory agreement is obtained between the third-law entropy of gaseous ferrocene and the entropy calculated from spectroscopic data with a restricting barrier of 0.9 kcal mole^{-1} as well as with a zero barrier[107].

Crystalline ferrocene is disordered at room temperature, but the majority of the molecules are probably staggered[108]. There is a λ-point transition at 163.9 °K with an accompanying entropy change of $\Delta S = R \ln 1.89$ cal deg^{-1} [109]. Crystals are probably ordered below the transition point, but the structure has not yet been determined[110]. It has been noted, however, that of the 17 nonbridged ferrocene derivatives studied by X-ray diffraction, 15 have dihedral angles $\phi \lesssim 17°$ and 13 have $\phi \lesssim 10°$, where $\phi = 0°$ corresponds to an eclipsed and $\phi = 36°$ to a staggered conformation[111]. Hence the tendency of ferrocene derivatives to accept an eclipsed conformation in a crystalline environment seems clear.

Ruthenocene, $(h_5\text{–}C_5H_5)_2Ru$, is eclipsed in the crystalline[112] as well as in the gas phase[103]. The main structure parameters are listed in Table 2. Again the C–H bonds are bent out of the plane of the cyclopentadienyl rings. Least-squares refinement without shrinkage corrections gave $< C_5$,CH = 5.6(0.9)°. If it is assumed that the shrinkage correction has the same magnitude as in ferrocene, $< C_5$, C–H = 4.7(0.9)°. The vibrational amplitude of the M–C distance shows that the bonding is as firm as or firmer than in ferrocene. It is interesting to note that the C–C bond distances in ferrocene and ruthenocene are very similar and significantly longer than the C–C bonds in derivatives of main group elements like $(C_5H_5)_2Be$ or $(C_5H_5)_2Mg$. It seems that the strengthening of the metal-to-ring bonds is accompanied by a weakening of the C–C bonds within the rings.

Chromocene[36], $(h_5\text{–}C_5H_5)_2Cr$, and nickelocene[113], $(h_5\text{–}C_5H_5)_2Ni$, have respectively two electrons *less* and two electrons *more* than ferrocene. Both compounds have two unpaired electrons[105]. It seems reasonable to assume that chromocene has two bonding electrons less and nickelocene two antibonding electrons more than ferrocene. Both compounds consequently have considerably longer, $\Delta R = 0.1$ Å, and looser, $\Delta l = 0.02$ Å, metal-to-carbon bonds than ferrocene, the bonds in nickelocene being slightly longer and looser than those in chromocene. The C–C bonds in the two compounds are very similar and slightly shorter than in ferrocene.

For chromocene the best agreement with experimental data was obtained with an eclipsed model, but a staggered conformation could not be ruled out. Probably the equilibrium conformation is eclipsed, but the barrier to rotation lower than in ferrocene. For nickelocene it was stated that staggered and eclipsed models gave equally good agreement.

Least- squares refinement including shrinkage corrections gave $< C_5$, C–H = 2.9(1.1)° in chromocene; the deviation from planarity is significant at the 2% level. The structure of nickelocene was refined without shrinkage corrections. The resulting angle $< C_5$, C–H = 0.3(1.5)° would become –0.6° if it is assumed that the shrinkage correction has the same magnitude as in ferrocene. In either case it is not significantly different from zero.

Manganese cyclopentadienide[114], $(h_5{-}C_5H_5)_2Mn$, has only one electron less than ferrocene, but since it has five unpaired electrons[105], it may be assumed that it has three bonding electrons less and two antibonding electrons more. This view is certainly in harmony with the molecular structure, since the M–C bond distance is more than 0.3 Å longer and the M–C vibrational amplitude more than twice as large as in ferrocene. Equally good agreement with experimental data is obtained with eclipsed and staggered models. Probably the barrier to internal rotation is of the order of 100 cal mole^{-1} or less. In this compound, in contrast to those discussed above, the M···H distance gives no isolated peak in the radial distribution curve, and consequently the angle $< C_5$, C–H cannot be determined with confidence.

8. References

1) Davis, M.: Electron diffraction in gases. New York: Marcel Dekker 1971.
2) Bauer, S. H., in: Physical chemistry, an advanced treatise (eds. H. Eyring, D. Henderson, and W. Jost), Vol. 4., Chap. 4. New York: Academic Press 1970.
3) Bartell, L. S., in: Physical methods in chemistry (eds. A. Weinberger and B. W. Rossiter). New York: Wiley-Interscience 1972.
4) Karle, J., in: Determination of organic structures by physical methods (eds. F. C. Nachod and J. J. Zuckerman), Vol. 5. New York: Academic Press 1972.
5) Seip, H. M., in: Molecular structure by diffraction methods (eds. G. A. Sim and L. E. Sutton), Vol. 1, p. 7. London: The Chemical Society 1973.
6) Robiette, G. A., in: Molecular structure by diffraction methods (eds. G. A. Sim and L. E. Sutton), Vol. 1, p. 160. London: The Chemical Society 1973.
7) Beagley, B., in: Molecular structure by diffraction methods (eds. G. A. Sim and L. E. Sutton), Vol. 1, p. 111. London: The Chemical Society 1973.
8) Gillespie, R. J.: J. Chem. Ed. *47*, 18 (1970).
9) Coates, G. E., Green, M. L. H., Wade, K.: Organometallic compounds. London: Menthuen & Co. Ltd. 1967.
10) Fetter, N. R.: Organometal. Chem. Rev. A *3*, 1 (1968).
11) Coates, G. E., Morgan, G. L., in: Advances in organometallic compounds (eds. F. G. A. Stone and R. West), Vol. 9, p. 195. New York: Academic Press 1970.
12) Snow, A. I., Rundle, R. E.: Acta Cryst. *4*, 348 (1951).
13) Rundle, R. E., Sturdivant, J. H.: J. Am. Chem. Soc. *69*, 1561 (1947).

14) Cowan, D. O., Krieghoff, N. G., Donnay, G.: Acta Cryst. *B24*, 287 (1968).
15) Spiro, T. G., Templeton, D. H., Zalkin, A.: Inorg. Chem. *7*, 2165 (1968).
16) Pauling, L.: The nature of the chemical bond, 3rd ed. Ithaca: Cornell University Press 1960.
17) Baird, N. C., Barr, R. F., Datta, R. K.: J. Organometal. Chem. *59*, 65 (1973).
18) Coates, G. E., Roberts, P. D., Downs, A. J.: J. Chem. Soc. *1967A*, 1085.
19) Almenningen, A., Haaland, A., Nilsson, J. E.: Acta Chem. Scand. *22*, 972 (1968).
20) Kovar, R. A., Morgan, G. L.: Inorg. Chem. *8*, 1099 (1969).
21) Almenningen, A., Haaland, A., Morgan, G. L.: Acta Chem. Scand. *23*, 2921 (1969).
22) Brendhaugen, K., Haaland, A.: Unpublished observation.
23) Atwood, J. L., Whitt, C. D.: J. Organometal. Chem. *32*, 17 (1971).
24) Drew, D. A., Haaland, A.: Acta Chem. Scand. *26*, 3079 (1972).
25) Drew, D. A., Haaland, A.: Acta Chem. Scand. *26*, 3351 (1972).
26) Akishin, P. A., Spiridonov, V. P.: Kristallografiya *2*, 475 (1957).
27) Almenningen, A., Bastiansen, O., Haaland, A.: J. Chem. Phys. *40*, 3434 (1964).
28) Haaland, A.: Acta Chem. Scand. *22*, 3030 (1968).
29) Fischer, E. O., Hoffmann, H. P.: Chem. Ber. *92*, 482 (1959).
30) Morgan, G. L., McVicker, G. B.: J. Am. Chem. Soc. *90*, 2789 (1968).
31) Ionov, S. P., Ionova, G. V.: Izv. Acad. Nank SSSR. Ser Khim. *1970*, 2836; Bull. Acad. Science USSR, Ser. Chem. *1970*, 2678.
32) McVicker, G. B., Morgan, G. L.: Spectrochimica Acta *26A*, 23 (1970).
33) Wong, C., Tseng-yuh, L., Kuei-jung, C., Lee, S.: Acta Cryst. *B28*, 1662 (1972).
34) Wong, C., Lee, T. J., Lee, T. W., Chang, T. W., Liu, C. S.: Inorg. Nucl. Chem. Letters, *9*, 667 (1973).
35) Drew, D. A., Haaland, A.: Acta Cryst. *B28*, 3671 (1972).
36) Haaland, A., Lusztyk, J., Novak, D. P., Brunvoll, J., Starowieyski, K. B.: J. C. S. Chem. Commun. *1974*, 54.
37) Bartell, L. S., Carroll, B. L.: J. Chem. Phys. *42*, 1135 (1965).
38) Carroll, B. L., Bartell, L. S.: Inorg. Chem. *7*, 219 (1968).
39) Bartell, L. S., Carroll, B. L.: J. Chem. Phys. *42*, 3076 (1965).
40) Eberhardt, W. H., Crawford, B. L., Lipscomb, W. N.: J. Chem. Phys. *22*, 989 (1954).
41) Lipscomb, W. N.: Boron hydrides. New York: W. A. Benjamin 1967.
42) Switkes, E., Stevens, R. M., Lipscomb, W. N.: J. Chem. Phys. *51*, 2085 (1969).
43) Turley, J. W., Rinn, H. W.: Inorg. Chem. *8*, 18 (1969).
44) Mole, T., Jeffery, E. A.: Organoaluminium compounds. Amsterdam: Elsevier 1972.
45) Henrickson, C. H., Eyman, D. P.: Inorg. Chem. *6*, 1461 (1967).
46) Lewis, P. H., Rundle, R. E.: J. Chem. Phys. *21*, 986 (1953).
47) Huffman, J. C., Streib, W. E.: Chem. Commun. *1971*, 911.
48) Almenningen, A., Halvorsen, S., Haaland, A.: Acta Chem. Scand. *25*, 1937 (1971).
49) Tables of interatomic distances and configuration in molecules and ions (ed. L. E. Sutton). London: The Chemical Society 1958.
50) Levison, K. A., Perkins, P. G.: Theoret. Chim. Acta *17*, 15 (1970).
51) Ramey, K. C., O'Brian, J. F., Hasegawa, I., Borchert, A. E.: J. Phys. Chem. *69*, 3418 (1965).
52) Williams, K. C., Brown, T. L.: J. Am. Chem. Soc. *88*, 5460 (1966).
53) Smith, M. B.: J. Organometal. Chem. *46*, 31 (1972).
54) Smith, M. B.: J. Phys. Chem. *71*, 364 (1967).
55) Smith, M. B.: J. Organometal. Chem. *22*, 273 (1970).
56) Wartik, T., Schlessinger, H. I.: J. Am. Chem. Soc. *75*, 835 (1953).
57) Almenningen, A., Anderson, G. A., Forgaard, F. R., Haaland, A.: Acta Chem. Scand. *26*, 2315 (1972).
58) Almenningen, A., Bartke, T. C., Haaland, A., Stokkeland, O.: Unpublished results.
59) Almenningen, A., Gundersen, G., Haugen, T., Haaland, A.: Acta Chem. Scand. *26*, 3928 (1972).
60) Hoffmann, E. G., Schomburg, W.: Z. Electrochem. *61*, 1101 (1957).
61) Kroll, W. R., Naegele, W.: Chem. Commun. *1969*, 246.
62) Haaland, A., Weidlein, J.: J. Organometal. Chem. *40*, 29 (1972).
63) Drew, D. A., Haaland, A.: Acta Chem. Scand. *27*, 3735 (1973).
64) Einstein, F. W. B., Gilbert, M. M., Tuck, D. G.: Inorg. Chem. *11*, 2832 (1972).
65) Anderson, G. E., Forgaard, F. R., Haaland, A.: Acta Chem. Scand. *26*, 1947 (1972).
66) Beagley, B., Hewitt, T. G.: Trans. Faraday Soc. *64*, 2561 (1968).
67) Gerteis, R. L., Dickerson, R. E., Lrown, T. L.: Inorg. Chem. *3*, 872 (1964).
68) McLean, W. J., Jeffrey, G. A.: J. Chem. Phys. *47*, 414 (1967).
69) Durig, J. R., Li, Y. S., Odom, J. D.: J. Mol. Struct. *16*, 443 (1973).
70) Bryan, P. S., Kuczkowski, R. L.: Inorg. Chem. *11*, 553 (1972).

71) Bryan, P. S., Kuczkowski, R. L.: J. Chem. Phys. *55*, 3049 (1971).
72) Hollander, F. J., Templeton, D. H., Zalkin, A.: Inorg. Chem. *12*, 2262 (1973).
73) Almenningen, A., Haaland, A., Haugen, T., Novak, D. P.: Acta Chem. Scand. *27*, 1821 (1973).
74) Grant, D. F., Killean, R. C. G., Lawrence, J. L.: Acta Cryst. *B25*, 377 (1969).
75) Gropen, O., Haaland, A.: Acta Chem. Scand. *27*, 521 (1973).
76) Zasorin, E. Z., Rambidi, N. S.: J. Structural Chem. *8*, 347 (1967).
77) Le Carpentier, J.-M., Weiss, R.: Acta Cryst. *B28*, 1437 (1972).
78) Laubengayer, A. W., Lengnick, G. F.: Inorg. Chem. *5*, 503 (1966).
79) Weidlein, J., Krieg, V.: J. Organometal. Chem. *11*, 9 (1968).
80) Davidson, N., Brown, H. C.: J. Am. Chem. Soc. *64*, 316 (1942).
81) Jeffrey, E. A., Mole, T.: J. Austr. J. Chem. *21*, 2683 (1968).
82) McLaughlin, G. M., Sim, G. A., Smith, J. D.: J. C. S. Dalton *1972*, 2197.
83) Krieg, V., Weidlein, J.: J. Organometal. Chem. *21*, 281 (1970).
84) Schmidbaur, H., Schindler, F.: Chem. Ber. *99*, 2178 (1966).
85) Drew, D. A., Haaland, A., Weidlein, J.: Z. anorg. allg. Chem. *398*, 241 (1973).
86) Gundersen, G., Haugen, T., Haaland, A.: Organometal. Chem. *54*, 77 (1972).
87) Ebsworth, E. A. V., in: Organometallic compounds of the group IV elements (ed. A. G. MacDiarmid), Vol. 1, Chap. 1. New York: Marcel Dekker 1968.
88) Mason, R., Mingos, D. M. P.: J. Organometal. Chem. *50*, 53 (1973).
89) Haaland, A., Stokkeland, O.: Unpublished results.
90) Brendhaugen, K., Haaland, A., Novak, D. P.: Acta Chem. Scand. *A28*, 45 (1974).
91) Brauer, D. J., Stucky, G. D.: J. Am. Chem. Soc. *91*, 5462 (1969).
92) Haaland, A., Stokkeland, O., Weidlein, J.: Preliminary results.
93) Shibata, S., Bartell, L. S., Gavin, R. M.: J. Chem. Phys. *41*, 717 (1964).
94) Frasson, E., Menegus, F., Panattoni, C.: Nature *199*, 1087 (1963).
95) Almenningen, A., Haaland, A., Motzfeldt, T.: J. Organometal. Chem. *7*, 97 (1967).
96) Panattoni, C., Bambieri, G., Croatto, U.: Acta Cryst. *21*, 823 (1966).
97) Veniaminov, N. N., Ustynyuk, Y. A., Alekseev, N. V., Ronova, I. A., Struchkov, Y. T.: Organometal. Chem. *22*, 551 (1970).
98) Veniaminov, N. N., Ustynyuk, Y. A., Struchkov, Y. T., Alekseev, N. V., Ronova, I. A.: Zhur. strukt. Khim. *11*, 127 (1970), J. Struct. Chem. *11*, 111 (1970).
99) Veniaminov, N. N., Ustynyuk, Y. A., Alekseev, N. V., Ronova, I. A.: Doklady Akad. Nauk S.S.S.R. *199*, 346 (1971); Doklady Chemistry *199*, 577 (1971).
100) Sharpen, L. H., Laurie, V. W.: J. Chem. Phys. *43*, 2765 (1965).
101) Bentham, J. E., Rankin, D. W. H.: J. Organometal. Chem. *30*, C 54 (1971).
102) Bohn, R. K., Haaland, A.: J. Organometal. Chem. *5*, 470 (1966).
103) Haaland, A., Nilsson, J. E.: Acta Chem. Scand. *22*, 2653 (1968).
104) Schäfer, L., Brunvoll, J., Cyvin, S. J.: J. Mol. Struct. *11*, 459 (1972).
105) Wilkinson, G., Cotton, F. A., in: Progress in inorganic chemistry, Vol. 1 (ed. F. A. Cotton), New York: Interscience 1959.
106) Coutiere, M.-M., Demuynak, J., Veillard, A.: Theoret. chim. Acta *27*, 281 (1972).
107) Haaland, A.: Acta Chem. Scand. *23*, 1836 (1969).
108) Willis, B. T. M.: Acta Cryst. *13*, 1088 (1960), AERE Report (Harwell) R 3708 (1961).
109) Edwards, J. W., Kington, G. L., Mason, R.: Trans. Far. Soc. *56*, 660 (1960).
110) Hyams, I. J., Ron, A.: J. Chem. Phys. *59*, 3027 (1973).
111) Krukonis, A. P., Silverman, J., Yannoni, N. F.: Acta Cryst. *B28*, 987 (1972).
112) Hardgrave, G. L., Templeton, D. H., Acta Cryst. *12*, 28 (1959).
113) Hedberg, L., Hedberg, K.: J. Chem. Phys. *53*, 1228 (1970).
114) Almenningen, A., Haaland, A., Motzfeldt, T. in: Selected topics in structure chemistry, Oslo: Universitetsforlaget 1967.

Received February 15, 1974

Stereochemistry of Compounds Containing Bonds between Si, P, S, Cl and N or O

Dr. Lev Vilkov and Dr. L. S. Khaikin

Department of Chemistry, Moscow State University, USSR

Contents

Introduction 26
1. Molecules Containing SiN Bonds 27
   1.1. Three-Coordinate N Atom 28
   1.2. Two-Coordinate N Atom 31
2. Molecules Containing PN Bonds 33
   2.1. Three-Coordinate N Atom 33
   2.2. Two-Coordinate N Atom 38
3. Molecules Containing SN Bonds 39
   3.1. Three-Coordinate N Atom 40
   3.2. Two- and Uni-Coordinate N Atoms 42
4. Molecules Containing ClN Bonds 43
   4.1. Three-Coordinate N Atom 44
   4.2. Two-Coordinate N Atom 45
5. Molecules Containing SiO Bonds 46
6. Molecules Containing PO Bonds 49
   6.1. Two-Coordinate O Atom 49
   6.2. Uni-Coordinate O Atom 53
7. Molecules Containing SO Bonds 55
   7.1. Two-Coordinate O Atom 55
   7.2. Uni-Coordinate O Atom 56
8. Molecules Containing ClO Bonds 60
   8.1. Two-Coordinate O Atom 60
   8.2. Uni-Coordinate O Atom 61
Discussion 62
References 65

Introduction

According to current theories, the third-row elements listed in the title can participate in σ and π bonding with their vacant d orbitals, while the N and O atoms are known to contain lone pairs in their valency shells. So one expects this paper to be a discussion of stereochemical patterns in terms of p–d interactions, which should result in quite specific molecular geometry, differing from the geometry characteristic of similar compounds of second-row elements.

It is a relatively long time since the unique character of the stereochemistry of compounds containing bonds between the Si, P, S or Cl atoms and nitrogen or, in the first place, oxygen was recognized on the basis of numerous crystallographic studies on silicates, phosphates, sulfates, chlorates, and phosphazenes, also oxides, oxyhalides, etc.[1,2].

Since then, much more reliable structural data have been collected, thanks to progress in X-ray and gas-phase electron diffraction techniques and the availability of high-precision microwave spectroscopy data.

The stereochemistry of nitrogen and oxygen derivatives is a very promising area, since the bond configuration at N and O undergoes considerable changes, depending on the chemical nature of the substituents (in contrast to *e.g.* carbon derivatives). Thus, the bond configuration around trivalent three-coordinate nitrogen changes from pyramidal in NH_3, NMe_3, etc. to planar in $N(SiH_3)_3$. A two-coordinate nitrogen atom forms a rather bent structure in HN_3 (113°) and a linear structure in H_3SiNCO. Valence angles around oxygen undergo similar changes in passing from Me_2O (ca. 111°) to $(SiH_3)_2O$ (ca. 144°). An increase in valence angles is associated with a decrease in elemental-nitrogen and elemental-oxygen bond lengths, which then become less than those calculated with the Schomaker-Stevenson equation[3]:

$$r_{AB} = r_A + r_B - 0.09\,|x_A - x_B|,$$

where r_A and r_B are the covalent radii of the A and B atoms and x_A and x_B are the corresponding electronegativities. At first, only the qualitative aspects of the model involving vacant d orbitals were applied to the explanation of these effects.

A comprehensive paper by Cruickshank[4] demonstrated from symmetry considerations that p and d AO's can participate in additional $(p_\pi - d_\pi)$ bonding between the four-coordinate Si, P, S, Cl atoms and the N or O atoms, and suggested a correlation between bond length and bond order. In recent years a wealth of calculational results has become available. Bartell *et al.*[5] demonstrated a correlation between the EHMO populations and Cruickshank's $(p_\pi - d_\pi)$ bond orders, using PO and SO bonds as an example. However, MO calculations and Cruickshank's model yielded quite different relative contributions of the various d orbitals to π-bonding. Furthermore, they showed[5] that variation of the PO and SO bond distances can be rationalized with complete neglect of d-orbital contributions, though including the d orbitals gives a better correlation between MO populations and bond lengths. The participation of d AO's in the SiO bonding involves similar problems[6–9].

Ab initio SCF-MO calculations of H_2NPH_2 have been performed[10] to show that the nitrogen atom should possess planar configuration, whether the basis set includes d orbitals or not, because of the inductive transfer of charge density from PH_2 to NH_2. On the other hand, CNDO/2 calculations of some sulfur derivatives[11, 12] yield an adequate description of the substituent effects on the bond lengths and valence angles only when the d orbitals on sulfur are included. Thus, even briefly outlined, the problem of interatomic influences is clearly exceedingly involved. Among the problems associated with the participation of d AO's on Si, P, S, and Cl, that of energy balance is the most difficult to solve, since d AO's of free atoms are characterized by significant energy values[13, 14].

It is not the purpose of the following discussion of the experimental data to provide arguments in favor of any of the current theories, or to develop a new theory that would explain all the available data. Rather, we consider it our task to classify and generalize these data in terms of classic structural theory, and thus to reveal patterns in the variation of the principal geometrical parameters: bond lengths, valence angles and, as far as possible, configurations of the predominant rotational isomers. Data on free molecules provide the most valuable information about the nature of interatomic influences. Accordingly, we concentrate on the gas-phase electron-diffraction and microwave-spectroscopy data. Since, as a rule, the most pronounced changes in molecular geometry occur in the fragments comprising N or O and their nearest neighbors, the data on these changes are listed in the tables. When significant changes in bond lengths and bond configurations occur at Si, P, S and Cl, they are discussed separately.

The data cited are taken from the original papers and therefore sometimes lack conformity, especially in representation of experimental errors[a]. The latter are given by standard deviations in some papers, by two or three times this value in others, and sometimes have been calculated from more general considerations.

The gas-phase electron-diffraction technique furnishes effective rather than equilibrium geometric parameters. This should be borne in mind in discussions of data on molecules that undergo large-amplitude low-frequency vibrations. The effective conformations of such molecules may differ significantly from equilibrium conformations. Bastiansen[15] was the first to observe this phenomenon in studying the allene molecule, $H_2C{=}C{=}CH_2$, where the effective distance between the terminal carbon atoms proved somewhat shorter than twice the C=C bond distance. This shrinkage effect was explained by Morino[16] and is known as the Bastiansen-Morino effect[17]. Thus, linear molecules are characterized by nonlinear effective structures. In the case of more complicated molecules, effective configurations lose some symmetry elements characteristic of the equilibrium ones. Thus, C_s is reduced to C_1, C_{2v} to C_2, D_{2d} to D_2, etc. In view of these complications, carefully collating the experimental data may prove an intricate problem.

1. Molecules Containing SiN Bonds

The first convincing evidence of the specific nature of the interactions between the third-row elements and nitrogen was reported in 1955 by Hedberg[18], who has studied the $N(SiH_3)_3$ molecule.

[a] These are given by figures in brackets, in numbers of the last digit.

This study, which has since become a classic, demonstrates a significant increase in valence angles around the nitrogen atom (in fact, they increase up to the highest possible value) associated with a considerable shortening of the SiN bonds to less than the value estimated for the single bond length (1.80 Å[3]). These results are consistent with the idea of $(p_\pi - d_\pi)$ interactions. If treated in terms of such interactions, they make good chemical sense, since $N(SiH_3)_3$ is a very weak electron donor compared to NMe_3. Despite the importance of these findings, no further investigations have been undertaken in this area for a long time. It is only recently that quite a few data have been reported on molecules containing the $-\overset{|}{\underset{|}{Si}}-N<$ (and also $-\overset{|}{\underset{|}{Si}}-N=$) fragment. An analysis of these data reveals that the stereochemical characteristics of the SiN interactions are determined by changes in the nitrogen valency state and by the influence of the environment nearest the SiN bond.

1.1. Three-Coordinate N Atom

Recently, the $N(SiH_3)_3$ molecule has been subjected to a new electron-diffraction study[19], which fully confirmed Hedberg's results. This study was undertaken in view of the increased precision of the up-to-date sector-microphotometer method. Average valence angles at three-coordinate nitrogen (α_N) hardly differ from 120° in most other structures studied (Table 1). Such is the case with molecules containing one or several Si atoms bonded to the single nitrogen atom and with molecules where several nitrogen atoms surround the single silicon atom. The observed deviations from planarity of about 0.5° can be attributed to the Bastiansen-Morino effect.

Wider valence angles at N result in a more planar structure of $(Me_2SiNH)_3$ as compared to cyclohexane. The six-membered cycle in this compound is characterized by the NNN/NSiN dihedral angle of 20°.

However, it must be borne in mind that locating the single hydrogen atom attached to N is very difficult, because the corresponding interatomic distances make small contributions to the scattering. Thus, in the case of $HN(SiH_3)_2$, the best fit is obtained with a planar configuration at the nitrogen atom, although ±0.03 Å displacements of hydrogen from the SiNSi planes cannot be ruled out.

The H_3SiNMe_2 molecule provides the only example of a nonplanar configurations. In this case, the angle between the SiN bond and the CNC plane is 27.8° (2.0°), yet α_N was found to be 117.0°, against 110.6° in NMe_3[34]. It is noteworthy that H_3SiNMe_2 is a stronger donor than disilyl- and trisilylamines[35].

The SiN bond lengths vary within rather wide limits, from 1.65 to 1.74 Å. They seem to be determined mainly by the chemical nature of the adjacent atoms. Thus the SiN bond length decreases with the number of Si atoms bonded to N, *e.g.* from 1.736 to 1.726 and to 1.715 Å along the series $(H_3Si)_3N \rightarrow (H_3Si)_2NMe \rightarrow H_3SiNMe_2$.

However, in the case of N-silylpyrrole containing one silyl group attached to the ring nitrogen, the SiN bond length is greater than in H_3SiNMe_2 and the same as in $(H_3Si)_3N$. A similar elongation of the SiN bond, up to 1.737 Å, occurs in $(H_3Si)_2NBF_2$. The effect can be attributed to the influence of the BF_2 group, with boron containing

Table 1. Structural data for molecules containing Si–N< bonds[1])

Molecules, Symmetry of frame	r (SiN), in Å and α_N [2]), in degrees	Bond angles of N atom, deg.	Method	Reference
$(H_3Si)_3N$ (~ D_{3h})	1.738(20) 119.6	SiNSi 119.6(1.0)	ED	18)
$(H_3Si)_3N$ (~ D_{3h})	1.734(2) 119.7	SiNSi 119.7(0.1)	ED	19)
$(H_3Si)_2NH$	1.725(3) 120[3])	SiNSi 127.7(0.1)	ED	20)
$(Me_3Si)_2NH$ (C_2)	1.735(12) 120[3])	SiNSi 125.5(1.8)	ED	21)
$(Me_2SiNH)_3$ slightly nonplanar cycle	1.728(3)	SiNSi 126.5(1.0)	ED	22)
$(Me_2SiNH)_4$	1.78(ass.)	SiNSi 123(4)	ED	23)
$(H_3Si)_2NMe$ (C_{2v})	1.726(3) 120.0	SiNSi 125.4(0.4) SiNC 117.3(0.2)	ED	24)
$(H_3Si)_4N_2$ (~ D_{2d})	1.731(4) 120.0	SiNSi 129.5(0.7) SiNN 115.3	ED	25)
$[(Me_3Si)_2N]_2Be$ (D_{2d})	1.722(7) 120.0	SiNSi 129.2(0.7) SiNBe 115.35	ED	26)
$(H_3Si)_2NBF_2$ (~ C_{2v})	1.737(4) 120	SiNSi 123.9(0.3)	ED	27)
$Me_3SiN(H)Me$	1.72(3) 120[3])	SiNC 130(5)	ED	28)
H_3SiNMe_2 (C_s)	1.715(4) 117.0	SiNC 120.0(0.4) CNC 111.1(1.2)	ED	29)
H_3Si–N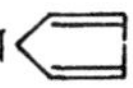 (C_{2v})	1.736(6) 120.0	SiNC 125.2(0.5) $(CNC)_{cycle}$ = = 109.6(0.9)	ED	30)
$H_3SiN(H)PF_2$ (C_1 and C_5)	1.720(8) 120[3])	SiNP 127.9(0.7)	ED	31)
F_3SiNMe_2 Staggered (ass.)	1.654(15) 120	SiNC 119.7(1.6) CNC 120.5(3.0)	ED	32)
Cl_3SiNMe_2 Staggered (ass.)	1.657(12) 119.8	SiNC 123.1(0.8) CNC 113.1(1.8)	ED	32)
$ClSi(NMe_2)_3$ (C_3)	1.715(4) 119.8	SiNC 120.5(0.5) CNC 118.5(1.5)	ED	33)

[1]) Here and below experimental errors are given in brackets, in numbers of the last digit.

[2]) α_N – mean value of bond angles of N atom.

[3]) The N–H bond position was not determined precisely.

a vacant p orbital. On the other hand, the electron-withdrawing substituents (F, Cl) attached to silicon cause shortening of the SiN bond, which becomes 1.654 Å in F_3SiNMe_2 and 1.657 Å in Cl_3SiNMe_2. However, the effect is most pronounced when silicon carries three halogeno and one amino substituents. With the inverse ratio of the substituents, *e.g.* in $ClSi(NMe_2)_3$, the shortening appears to be negligible or nonexistent. In all cases, however, the SiN bond length is considerably less than the Schomaker-Stevenson estimate of 1.80 Å.

It is noteworthy that the GeN bond distance in $(H_3Ge)_3N$ (1.836 ± 0.005 Å)[36] is larger than the Schomaker-Stevenson estimate (1.81 Å) despite its planar structure. The $(H_3Si)_3P$ molecule has a rather nonplanar frame (∠ SiPSi = 96.5 ± 0.5°)[37]. However, in this case the PSi bond length (2.248 ± 0.003 Å[37]) approaches the single bond length estimate of 2.24 Å. The pyramidal configuration around the P atom seems to be predetermined by high barriers to inversion (27.4 kcal/mole in PH_3 against 5.9 kcal/mole in NH_3[38]).

Barriers to rotation about the SiN bond should be relatively low since molecular conformations are mainly determined by interactions between nonbonded atoms. By reason of such interactions, the CNC planes of dimethyl amino groups are turned so as to make a 67.5 (3.5)° dihedral angle with the corresponding ClSiN planes in $(Me_2N)_3SiCl$. Similar effects seem to be responsible for a nearly D_{2d} configuration of $(H_3Si)_4N_2$ with the angle of rotation about the NN bond of 82.5 (0.8)°. An analogous conformation occurs in $[(Me_3Si)_2N]_2Be$ (rotation around the NBeN axis).

The amino function influences all the other bonds at the Si atom to a certain degree. The data on chloroderivatives of silamines (Table 2) provide the best demonstration of these effects. The SiCl bonds in silamines undergo elongation associated with shortening of the SiN bonds. Again, the most pronounced effect is observed where there is only one SiCl bond in the molecule. In $(Me_2N)_3SiCl$ the SiCl bond is 0.03 Å longer than the Schomaker-Stevenson estimate of 2.05 Å. Conversely, the SiCl bond becomes about 0.09 Å shorter in the presence of stronger electron acceptors, *e.g.* F in F_3SiCl.

Table 2. SiCl bond lengths in some silicon derivatives

Molecule	*r* (SiCl), Å	Method of determination	References
$SiCl_4$	2.015(2)	ED	39)
$HSiCl_3$	2.021(2)	MW	40)
Me_2NSiCl_3	2.023(5)	ED	32)
$(Cl_3Si)_2O$	2.011(4)	ED	41)
H_3SiCl	2.0479(7)	MW	42)
$(Me_2N)_3SiCl$	2.082(6)	ED	33)
F_3SiCl	1.989(18)	MW	43)
$Cl_3SiN{=}C{=}O$	2.014(5)	ED	44)
$Cl_2Si(N{=}C{=}O)_2$	2.024(5)	ED	44)
$ClSi(N{=}C{=}O)_3$	2.020(9)	ED	44)

1.2. Two-Coordinate N Atom

Gas-phase studies of the simplest compounds of the types R–NCO, R–NCS and R–N=N≡N, where R is H or CH_3, yield values of valence angles at the –N= atom that are considerably less than 180°, thus indicating rather bent structures (Table 3). Re-

Table 3. Valence angles at N in the simplest pseudohalides

Molecule	RNY angle, degrees, (Y=C, N)	Method of determination	References
HN=C=O	128.1(0.5)	MW	[45]
HN=C=S	131.25(0.25)	MW	[46]
HN=N≡N	112.6(0.5)	MW	[47]
MeN=C=O	140.3(0.4)	ED	[48]
MeN=C=S	141.6(0.4)	ED	[48]
MeN=N≡N	116.8(0.3)	ED	[48]

placement of H or Me by a silyl group results in a significant increase of the angle. Thus, the H_3SiNCO and H_3SiNCS molecules possess C_{3v} symmetry according to the microwave data (Table 4).

Electron diffraction studies of H_3SiNCO and H_3SiNCS also yield effective values for the SiN=C angle about 20° larger than those (HN=C, CN=C) characteristic of the simplest pseudohalides. However, the equilibrium values could not be approximated by the electron-diffraction technique because of the large-amplitude, low-frequency bending vibrations of the Si–N=C group, which is responsible for considerable populations of higher vibrational levels. According to the authors of[50], the potential surface of such vibrations may have one (linear configuration) or two minima with a low energy barrier.

The microwave experiments indicate a linear configuration of the SiN=C=X groups in the ground vibrational state. Shrinkage corrections improve the agreement between electron-diffraction and microwave data.

A similar increase in the valence angles at N has been observed in the F and Me derivatives, F_3SiNCX and Me_3SiNCX (X = O, S), by the electron-diffraction technique. In this connection it is worth mentioning that the valence angle at N in $Si(NCO)_4$ and its Cl-substituted derivatives is essentially the same as in MeNCO.

The Me_3Si group effects some increase (however, only to ca. 130°) of the Si–N=N angle in Me_3SiN_3 as compared to the C–N=N angle in Me–N=N≡N. Comparing the data on the germanium derivatives H_3GeNCO and H_3GeN_3 reveals a similar decrease in the valence angle at N on passing from isocyanate (141.6°) to azide (119°)[58].

Anderson, Rankin and Robertson[48] analyzed valence angles in azides and isocyanates in terms of the following schemes:

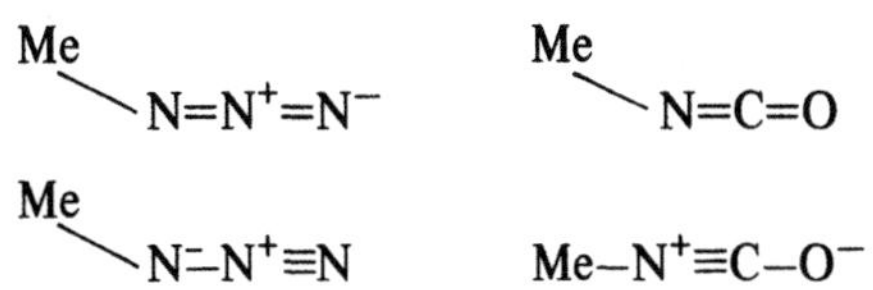

Table 4. Structural data for molecules containing Si–N= bonds

Molecules and symmetry of frame	Si–N= bonds in Å	Si–N=Y angles (Y=C, N) in degrees	Method	References
$H_3Si–N=C=O$ (C_{3v})	1.699(10)	180	MW	49)
$H_3Si–N=C=O$ (C_s)	1.703(4)	152.7(1.2)	ED	50)
$H_3Si–N=C=S$ (C_{3v})	1.714(10)	180	MW	51)
$H_3Si–N=C=S$ (C_s)	1.704(6)	163.8(2.6)	ED	50)
$H_3Si–N=N≡N$ (C_s)	–	< 180	MW	52)
$H_3Si–N=N≡N$ (C_s)	–	126	ED	53)
$Me_3Si–N=C=O$ (C_s, staggered)	1.76(2)	150(3)	ED	54)
$Me_3Si–N=C=S$ (C_s, staggered)	1.78(2)	154(2)	ED	54)
$Me_3Si–N=N≡N$ (C_1)[1]	1.734(7)	128.0(1.6)	ED	55)
$F_3Si–N=C=O$	1.648(10)	160.7(1.2)	ED	56)
$Cl_3Si–N=C=O$ (C_1)[2]	1.646(8)	138.0(0.4)	ED	44)
$Cl_2Si(–N=C=O)_2$ (C_2)[3]	1.687(4)	136.0(1.0)	ED	44)
$ClSi(–N=C=O)_3$ (C_3)[4]	1.684(5)	145(2)	ED	44)
$Si(–N=C=O)_4$ (T_d)	1.688(3)	146.4	ED	57)

[1]) Dihedral angle CSiNN 36.2(4.9)°.
[2]) Dihedral angle ClSiNC 24.0(4.0)°.
[3]) Dihedral angle ClSiNC 14.0(3.0)°.
[4]) Dihedral angle ClSiNC 85.0(6.0)°.

With azides, both schemes predict a bent structure, whereas in the case of isocyanates one scheme suggests a linear coordination. One may wonder whether such schemes can provide the basis for predicting geometric characteristics.

Distances between nonbonded atoms increase with the valence angle at N, so one can hardly expect high barriers to rotation about the Si–N bond. Thus, essentially free rotation occurs in Si $(NCO)_4$. Most other molecules do not possess high symmetry.

The $-\overset{|}{\underset{|}{Si}}-N=$ bond distance in pseudohalides varies still more widely (range 1.65 to 1.78 Å) than the $-\overset{|}{\underset{|}{Si}}-N\lt$ distance in aminosilanes. The upper limit approaches the estimate for single bond length (1.80 Å). There is no correlation between the SiN bond distances and effective SiN=C bond angles in pseudohalides. The SiN bond length decreases on passing from Me_3Si- to H_3Si-substituted isocyanates and further to halogeno-substituted derivatives. As for the SiN=C angle, it reaches its highest value in H_3Si derivatives. Note also that the shortest SiN bond occurs in a nonlinear molecule, $Cl_3SiN{=}C{=}O$, where the SiN=C angle is smallest. This result cannot be rationalized in terms of $(p_\pi - d_\pi)$ interactions. It appears that the role of Si d orbitals is not essential in this case. The effect of the N=C=O groups on the SiCl bond distance in chloroderivatives of isocyanates (2.020 ± 0.005 Å on the average) is negligible, although the SiCl bonding appreciably influences the SiN bond length. In contrast to the case of aminosilanes, in this case the SiCl bond distance remains constant to within experimental error (Table 2).

2. Molecules Containing PN Bonds

We have published two review papers[59, 60] which contain a more general survey of the stereochemistry of phosphorus. Here we concentrate on electron-diffraction and microwave data abstracted from these papers and from more recent original works.

2.1. Three-Coordinate N Atom

Like silicon, phosphorus effects a considerable widening of the amino group average valence angle at N, α_N (Table 5). However, in this case the widening appears to be less than that in silicon derivatives. Often the α_N parameter does not reach its limiting value of 120°. Phosphorus amino derivatives are subject to more pronounced effects due to the chemical nature of the substituents, especially at the nitrogen atom. Thus, in $P[N(CH_2)_2]_3$ where each of the nitrogen atoms forms a three-membered cycle, the α_N value is 101.5°, hence considerably less than the α_N value in $P(NMe_2)_3$ (117.5°). Probably, in the presence of halogeno substituents (F, Cl) at the P atom, the α_N value increases up to about 120°. As there is no agreement between the electron-diffraction and microwave data on H_2NPF_2 and Me_2NPF_2, one cannot reach a final conclusion as to the angle values. The microwave studies yield the α_N value of 120° for both molecules, whereas the electron diffraction results point to α_N values of 115.3 and 116.0°, respectively. The more recent microwave studies have been carried out since the electron-diffraction results were known; we believe that reinvestigation of the molecules by electron diffraction is needed to eliminate the discrepancy.

Table 5. Structural data for molecules containing P–N$<$ bonds

Molecules and symmetry of frame	P–N$<$ bonds, in Å α_N in degrees	Bond angles of N atom, in degrees	Method	References
$P[N(CH_2)_2]_3$ (C_3)	1.75(1) 101.5	PNC 121.0(1.5) CNC 61.5(0.5)	ED	61)
$P(NMe_2)_3$ (C_3)	1.700(5) 117.5	PNC 119.5(1.5) CNC 113.5(1.5)	ED	61)
$ClP(NMe_2)_2$ (C_1)	1.730(5) 120.0	PNC 119.8(0.5) CNC 120.3(0.7)	ED	62)
Me_2NPCl_2 (C_s, anti)	1.69(3) 120	PNC 120(2) CNC 120(2)	ED	63)
$Me_2NP(=O)Cl_2$ (C_1)	1.67(4) 116	PNC 116(2) CNC 116(2)	ED	63)
H_2NPF_2 (C_s, gauche)	1.650(4) 120.0	PNH 121.4 HNH 117.2(0.4)	MW	64)
H_2NPF_2 (C_1)	1.661(7) 115	PNH 119(2) HNH 108(3)	ED	65)
$H_3SiN(H)PF_2$ (C_1 and C_s)	1.657(7) 120	PNH 118.8 PNSi 127.9(0.7)	ED	31)
Me_2NPF_2 (C_s, gauche)	1.66 120	PNC 123 CNC 114	MW	67)
Me_2NPF_2 (C_1)	1.684(8) 116.1	PNC 118.3(0.6) CNC 111.8(1.5)	ED	65)
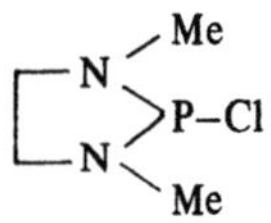 (C_s, envelope)	1.68 (ass.) 117	$(PNC)_{cycle}$ 113.7(1.0) PNC_{Me}= =CNC_{Me} 118.8(2.5)	ED	68)
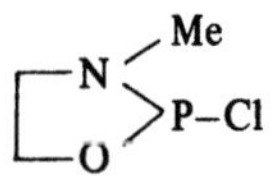 (C_1, envelope)	1.692(13) 120	$(PNC)_{cycle}$ 112.7(1.0) PNC_{Me} 125(2) CNC_{Me} 121(2)	ED	69)
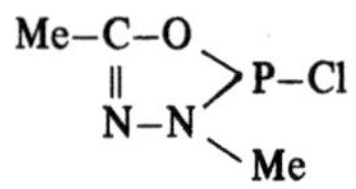 (C_1, envelope)	1.700(15) 120	$(PNN)_{cycle}$ 108	ED	70)
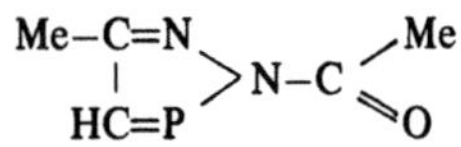 (planar cycle)	1.68(1) 120	$(PNN)_{cycle}$ 115.0(2.5)	ED	71)
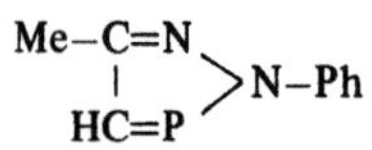 (planar cycle)	1.65(3) 114	$(PNN)_{cycle}$ 110 PNC_{Ph}= =NNC_{Ph} 116	ED	72)
$(F_3PNMe)_2$ (C_{2h}, planar cycle)	1.735 axial 1.595 equator. 120	$(PNP)_{cycle}$ 102.1 PNC_{Me} {134.3, 123.6}	ED	73)

Changes in the valency state of the phosphorus atom also play a certain role in determining molecular geometry. In contrast to Me_2NPCl_2, which contains three-coordinate phosphorus, α_N decreases to 116° in $Me_2NP(=O)Cl_2$, where the phosphorus atom is four-coordinate. When bonded to the five-coordinate ring phosphorus as in $(F_3PNMe)_2$, the ring nitrogen atom adopts a planar configuration.

As in the case of the SiN bonds, the PN distance values in phosphorus amino derivatives are less than the Schomaker-Stevenson estimate of 1.75 Å. In most structures the PN bond length is between 1.65 and 1.70 Å. The replacement of two amino groups by halogen atoms (F, Cl) in $P(NMe_2)_3$ results in the shortening of the only remaining PN bond; this is more pronounced with the fluorine substituents. This effect is similar to, though weaker than, the previously discussed effect observed in silicon derivatives. Thus, the SiN bond length decreases by about 0.06 Å on passing from $ClSi(NMe_2)_3$ to Cl_3SiNMe_2 or F_3SiNMe_2, whereas in the similar phosphorus derivatives ($P(NMe_2)_3 \rightarrow Cl_2PNMe_2$, F_2PNMe_2) the PN bond length decreases by only 0.01 und 0.03 Å, respectively.

The PN bond distance nearest to the Schomaker-Stevenson estimate occurs in two molecules containing three-coordinate phosphorus. However, one of them, $P[N(CH_2)_2]_3$, is characterized by α_N of only 101°, whereas the other, $ClP(NMe_2)_2$, possesses a planar configuration at the nitrogen atom with α_N of 120°.

We have pointed out[59)] that compounds containing four-coordinate phosphorus are characterized by shorter bonds and wider valence angles than the three-coordinate phosphorus derivatives. The shortening of the PN bond on passing from Me_2NPCl_2 to $Me_2NP(=O)Cl_2$ serves an example. This shortening is associated with a decrease of α_N to 116°, as mentioned already.

Planar five-membered ring structures of phosphadiazoles[71, 72)] were among the first known structures incorporating two-coordinate phosphorus[b)]. The ring PN bonds in these compounds appear to be somewhat shorter than those in nonplanar five-membered rings containing three-coordinate phosphorus[69, 70)].

The dimeric cyclophosphazene molecule $(F_3PNMe)_2$ contains distorted trigonal bipyramids around each of the phosphorus atoms, with an axial-equatorial orientation of the planar four-membered cycle with respect to the bipyramids. Thus, in this case the rule does not hold that fluorine atoms should be in axial positions in trigonal bipyramids of substituted fluorophosphoranes[80, 81)]. In such a structure the PN bonds are not equivalent, with axial bond distances approximating the Schomaker-Stevenson estimate of 1.75 Å. The equatorial bond distances, 1.595 Å, are considerably smaller than all the other PN bond distances we are aware of.

It has been mentioned already that the PN bonds of halogeno-substituted aminophosphines undergo less shortening than SiN bonds in the corresponding silamines. However, the effect of amino function is much more pronounced in the case of the former compounds and results in a greater elongation of PCl than of SiCl bonds.

The PCl bond in Me_2NPCl_2 is 0.04 Å longer than that in PCl_3 (Table 6) and 0.07 Å longer than the Schomaker-Stevenson estimate, yet only 0.01 Å less than the

b) Two-coordinate phosphorus also occurs in phosphamethinecyanine cations[74–76)] and phosphabenzenes[77–79)], where phosphorus replaces one of the carbon atoms in the benzene ring.

Table 6. PCl bond distances

Molecule	r(PCl), Å	Method	References
PCl_3	2.043(3)	MW	82)
PCl_3	2.042(2)	ED	83)
Me_2NPCl_2	2.083(5)	ED	63)
$(Me_2N)_2PCl$	2.180(4)	ED	62)
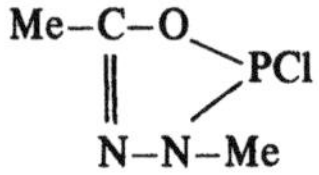	2.170(5)	ED	70)
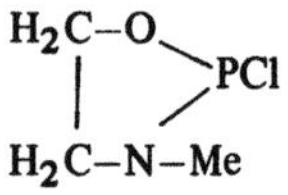	2.174(9)	ED	69)
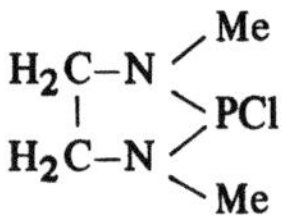	2.19(2)	ED	68)
F_2PCl	2.030(6)	MW	84)
$POCl_3$	1.989(2)	MW	85)
$POCl_3$	1.993(3)	ED	86)
$Me_2NP(=O)Cl_2$	2.033(8)	ED	63)
$Cl_2P(=O)N=C=O$	2.006(5)	ED	87)
$(NPCl_2)_3$	2.006(3)	ED	88)

sum of the covalent radii. Similarly, the PCl bond in $Me_2NP(=O)Cl_2$ is 0.04 Å longer than in $POCl_3$.

Our studies[70] on the Me–N–C(H)=C(Me)–O–P–Cl molecule provided the first example of a PCl bond that is considerably longer than the sum of the covalent radii. Similar phenomena were later observed by Naumov[62, 68, 69]. Thus, the PCl bond in $(Me_2N)_2PCl$ is as much as 0.14 Å longer than the PCl bonds in PCl_3 and 0.10 Å – longer than the sum of the covalent radii. The SiCl bond distances vary within far narrower limits.

The AB bond shortening to less than the sum of the covalent radii is often treated in terms of increasing bond order. According to the Schomaker-Stevenson equation, the polarity of the bond also contributes to the bond shortening. A formal interpretation of the observed PCl bond elongation in aminophosphines would involve bond orders of less than unity; this points to the necessity of reconsidering the generally accepted concepts on bond order.

There are not enough data on conformations of phosphorus amino derivatives to permit a conclusive decision as to what factors determine rotational isomerism about the PN bond. Strong interactions between nonbonded atoms in $P(NMe_2)_3$ and $P[N(CH_2)_2]_3$ should play an important role in determining their configurations, yet the actual rotamers (Fig. 1a and b) are not sterically favorable. The shortest distances between atoms in neighboring amino groups are 2.57 (C . . . C) and 2.78 (C . . . N) Å in $P(NMe_2)_3$ and 3.48 (C . . . C) and 2.80 (C . . . N) Å in $P[N(CH_2)_2]_3$.

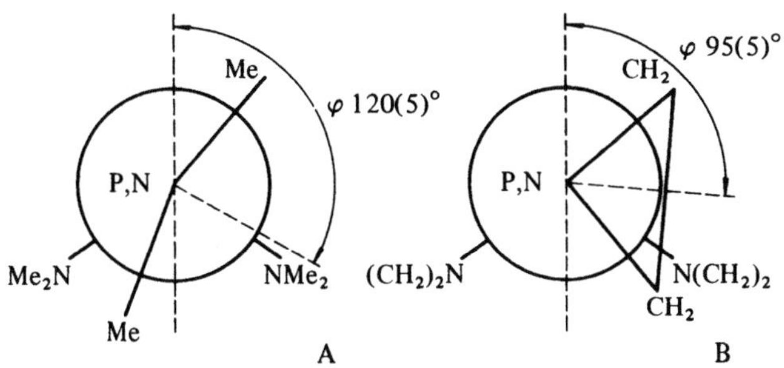

Fig. 1. Projections along a PN bond for tris-(dimethylamino)phosphin (a) and tris-(ethyleneimino)-phosphin (b)

There is some disagreement as to the conformational preference of R_2NPX_2 molecules[59]. The electron diffraction data are consistent with an "anti" conformation of Me_2NPCl_2 (C_s symmetry) whereas spectral data[89] favour the "gauche" form (C_s) (Fig. 2a and b). On the other hand in molecules of the type

$$\text{Me–}\overline{\text{N–CH}_2\text{–CH}_2\text{–N(Me)–P}}\text{–Cl,}$$

the PCl bonds occur only in axial orientation with respect to the five-membered envelope cycle, *i.e.* they are gauche with respect to the NC bonds. In the $(Me_2N)_2PCl$ molecule, the two C_2N planes form different dihedral angles, 76 and 156°, with the NPN plane.

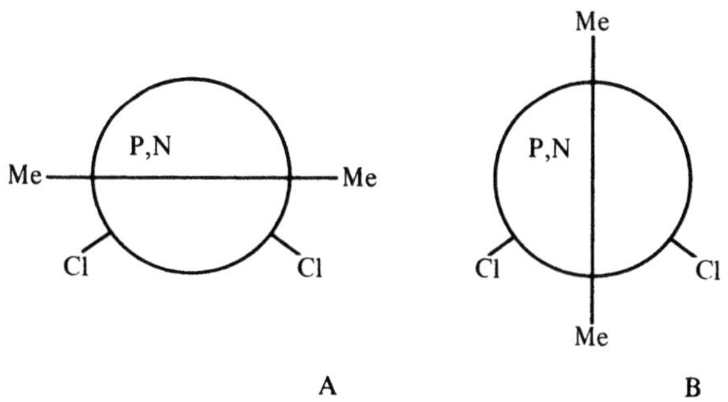

Fig. 2. Projections along a PN bond for "anti" (a) and "gauche" (b) molecular models of Me_2NPCl_2

The $Me_2NP(=O)Cl_2$ molecule (Fig. 3) has no symmetry elements. In the case of this molecule, the factors determining the conformational type may involve attractive interactions of the Me groups with the Cl atoms or with the polar P=O bond. Such interactions have been utilized in explaining the gauche molecular geometry of $CH_3CH_2CH_2Cl$[90] and CH_3OCH_2Cl[91].

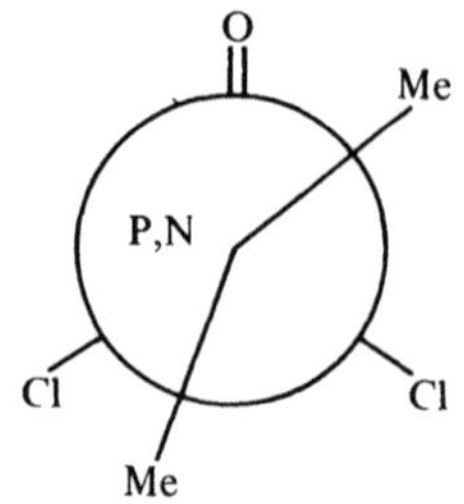

Fig. 3. Projection along a PN bond for $Me_2NP(=O)Cl_2$

2.2. Two-Coordinate N Atom

The P–N=C valence angles in the three-coordinate phosphorus pseudohalides (Table 7) are somewhat narrower than the valence angle at N in MeN=C=O (140.3°). The four-coordinate phosphorus atom in $Cl_2P(=O)N=C=O$ effects a further decrease of the P–N=C valence angle to 120°. As in the case of amino derivatives, one can again see that phosphorus and silicon give rise to rather different stereochemical effects (see Table 4). It appears to be a general rule that valence angles at N bonded to P are narrower than the corresponding angles in silicon compounds.

The PN bond length is practically the same (1.68 Å) in all the phosphorus pseudohalides studied. It hardly differs from the PN bond length in halogeno-substituted

Table 7. Structural data for molecules containing P–N= or P⋯N⋯ bonds

Molecules and symmetry of frame	r (P–N=) or r (P⋯N⋯), in Å	P–N=C or P⋯N⋯P angles, in degrees	Method	References
$F_2PN=C=O$ ($\sim C_s$)	1.683(6)	130.6(0.8)	ED	92)
$F_2PN=C=S$ ($\sim C_s$)	1.686(7)	140.5(0.7)	ED	92)
$(F_2P-N=)_2C$ (C_2)	1.680(6)	132.8(0.5)	ED	93)
$Cl_2P(=O)N=C=O$ (C_s)	1.684(10)	120.0(1.5)	ED	87)
$(NPCl_2)_3$ (C_{3v})	1.582(5)	119.7(0.2)	ED	88)
$(NPF_2)_3$ (D_{3h}, planar cycle)	1.586(13)	120	ED	94)
$(NPF_2)_4$	1.56		ED	95)

aminophosphines and is only 0.02 Å less than that in $P(NMe_2)_3$ (see Table 5). However, there are no data that indicate what influence electron-withdrawing substituents (F, Cl) exert on the two-coordinate nitrogen-to-phosphorus bonding.

The PCl bond distance in $Cl_2P(=O)N=C=O$ is the same as that in $POCl_3$ to within experimental error, while in $Me_2NP(=O)Cl_2$ the bond is longer by 0.04 Å (see Table 6). Similar trends are observed in silicon compounds containing two- and three-coordinate nitrogen.

The equilibrium configuration of $F_2PN=C=X$ (X = O, S, NPF_2) is probably that with "trans" arrangement of the PF_2 group relative to the N=C bond (C_s symmetry). The observed deviations of the effective structure from this configuration do not exceed 15°. It is interesting that the allene-like –N=C=N– system proves nonrigid as well. The effective dihedral angle in PN=C=NP amounts to 55° with respect to the "cis" arrangement of the NP bonds.

There appears to be some similarity in the stereochemistry of phosphorus pseudohalides and cyclophosphazenes. The latter may be described formally as cyclic structures containing a varying number of identical units $-\overset{|}{\underset{|}{P}}=N-$ or $=\overset{|}{\underset{|}{P}}-N=$.

In the first place, the valence angle at the nitrogen atom in the both types of compounds never exceeds 150°, though wider angles might be expected in higher-membered phosphazene cycles from purely geometrical considerations. A valence angle at nitrogen of 120° is observed in the $(NPCl_2)_3$ and $(NPF_2)_3$ planar trimeric molecules studied by the electron-diffraction technique. According to the X-ray data, the PNP angle reaches its optimum value of about 130 to 135° in tetrameric species which possess sufficiently nonplanar structures. In addition, phosphazenes and phosphorus pseudohalides are similar in that the PCl bond lengths are the same in $(NPCl_2)_3$ and $Cl_2P(=O)N=C=O$ (2.006 (4) Å), but considerably less than the PCl bond length in the amino derivative $Me_2NP(=O)Cl_2$ (see Table 6).

However, the PN bonds in phosphazenes are considerably shorter than those observed in both amino derivatives and phosphorus pseudohalides, presumably because there is no actual alternating of single and double bonds in the formally unsaturated $(NPX_2)_n$ ring systems. These bonds are characterized by marked sensitivity to the influence of the substituents at the phosphorus atom[60)].

3. Molecules Containing SN Bonds

Sulfur occurs in three formally positive oxidation states, usually designated by the roman numerals II, IV and VI. The formation of multiple bonds (especially with oxygen) is characteristic of sulfur and gives rise to a variety of stereochemical types. Sulfur can have 11 different valency states in nonionized structures under normal conditions. Unfortunately, there are relatively few gas-phase studies of molecules containing the SN bond, and this complicates the analysis of sulfur stereochemistry.

Table 8. Structural data for molecules containing S–N< bonds

Molecules, symmetry of frame	r (SN), in Å and α_N, in degrees.	Bond angles of N atom, degrees	Method	References
$S(NMe_2)_2$ (C_{2v})	1.688(6) 117.4	SNC 117.9(0.6) CNC 116.5(1.6)	ED	96)
$SO(NMe_2)_2$ (C_1)	1.693(4) 115.4	SNC 116.1(0.5) CNC 113.9(1.5)	ED	97)
Me_2NSO_2Cl (C_s)	1.69(2) 112	SNC= =CNC 112(2)	ED	98)
$SO_2(NMe_2)_2$ (C_{2v})	1.651(3) 116.1	SNC 115.2(1.1) CNC 118.0(3.2)	ED	99)
F_5SNF_2 (C_s)	1.697(4) 107.0	SNF 111.4(0.3) FNF 98.1(0.8)	ED	100)

3.1. Three-Coordinate N Atom

Sulfur amino derivatives provide no examples of planar configuration at N (Table 8). However, as a rule the α_N value in these molecules is far greater than the tetrahedral angle value of 109.5°. Solely in the case of F_5SNF_2, the α_N angle is 107°. Such a low value, however, may be regarded as the result of the fluorine substituent effect, as observed in other fluoroamines.

The available data are insufficient to reveal patterns of α_N in relation to the valency state of the sulfur atom. Thus, α_N decreases from 117.1° to 115.4° on passing from $S(NMe_2)_2$ to $SO(NMe_2)_2$. However, there is no consistency between data on the bond configurations at N in Me_2NSO_2Cl and $SO_2(NMe_2)_2$, and besides, these data lack accuracy. The electron diffraction and X-ray[101)] data on $SO_2(NMe_2)_2$ do not agree either. The X-ray parameters are: r (SN) = 1.623 (5) Å; ∠ SNC = 118.8 (0.4)°; ∠ CNC = 112.9 (0.4)°. The difference between these values and the electron-diffraction parameters is beyond the limits of measuring errors. A further investigation of the structure apppears to be necessary.

The SN bond length is smaller than the Schomaker-Stevenson estimate (1.73 Å) by values of the same order of magnitude as in the case of the PN bond length. With the exception of $SO_2(NMe_2)_2$, r(SN) values vary within the measuring error limits. It is of interest that in $SO_2(NMe_2)_2$, the bond shortening (to ca. 1.651 Å as against ca. 1.69 in other compounds) does not result in a planar configuration at N. The replacement of one amino group by chlorine is probably associated with an elongation of the SN bond, though one cannot be quite sure about it in view of the large experimental errors.

A much more pronounced effect is that of the amino function on the SCl bond length in the Me_2NSO_2Cl molecule (Table 9). This bond is 0.065 Å longer than the sum of the covalent radii, 2.03 Å. In most cases, the SCl bond lengths are greater than the Schomaker-Stevenson estimate of 1.99 Å.

Table 9. SCl bond distances

Molecule	r (SCl), Å	Method of determination	References
SCl_2	2.006(4)	ED	102)
S_2Cl_2	2.057(2)	ED	103)
$SOCl_2$	2.076(6)	ED	104)
ClS≡N	2.159(3)	ED	105)
SO_2Cl_2	2.011(5)	ED	104)
$MeSO_2Cl$	2.046(4)	ED	106)
Me_2NSO_2Cl	2.094(10)	ED	98)

The SCl bond length in Me_2NSO_2Cl is about 0.085 Å longer than that in SO_2Cl_2, an elongation almost as great as that observed for PCl in the case of aminophosphines.

A considerable increase in the valence angle at S in $S(NMe_2)_2$ is somewhat unexpected. In this compound, the angle amounts to 114.5°(1.6°) against 98.9(0.2)° in SMe_2 [107] and 103.0(0.4)° in SCl_2 [102].

This effect is less pronounced in sulfuryl derivatives, SO_2X_2, and is not detectable in thionyl compounds, SOX_2 (X = NMe_2, Me, Cl)[96].

All the sulfamides studied show conformational preference for the staggered arrangement of bonds relative to the SN bond (Fig. 4).

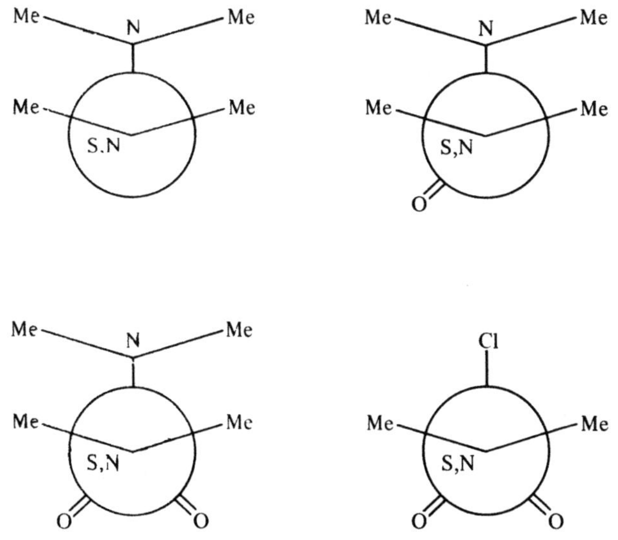

Fig. 4. Projections along a SN bond for $(Me_2N)_2S$, $(Me_2N)_2SO$, $(Me_2N)_2SO_2$ and Me_2NSO_2Cl

3.2. Two- and Uni-Coordinate N Atoms

In this section, for we are the first time discussing both single and multiple SN bonds (Table 10)[c]. In a five-membered cycle of thiadiazole, $H_2C_2N_2S$, the –S–N= bonds are 0.06 Å shorter than in $S(NMe_2)_2$ and about 0.1 Å shorter than the Schomaker-Stevenson estimate. These molecules are usually regarded as aromatic systems, which explains

Table 10. Structural data for molecules containing S–N=, S=N– or S≡N bonds

Molecules, symmetry of frame	SN bond lengths in Å	Bond angle of N atom in degrees	Method	References
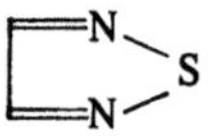 (C_{2v})	1.632(5)	106.5(0.4)	ED	108)
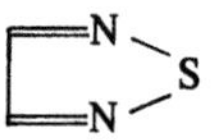 (C_{2v})	1.631(3)	99.55(0.2)	MW	109)
FN=S (C_s)	1.62(3)	122(3)	ED	110)
HN=S=O (C_s)	1.512(5)	115.8(1.0)	MW	111)
ClN=S=O (C_1)	1.559(4)	116.3(0.4)	ED	112)
$ClN{=}SF_2$ (C_s)	1.476(3)	120.0(0.2)	ED	113)
$ClN{=}SOF_2$ (C_s)	1.488(12)	115.0(0.7)	ED	114)
$Me_2S({=}O){=}NH$ (C_s)	1.521(3)	114.1(2.4)	ED	115)
$Me_2S({=}NH)_2$ (C_{2v})	1.533(2)	114.1(2.4)	ED	116)
ClS≡N (C_s)	1.448(3)	–	ED	105)
FS≡N (C_s)	1.446(10)	–	MW	117)
$F_3S{\equiv}N$ (C_{3v})	1.416(3)	–	MW	118)

[c] The formation of multiple bonds (in the sense of classic theory) is not characteristic of silicon compounds. Though phosphorus compounds containing multiple $-\overset{|}{\underset{|}{P}}=N-$ bonds are known, their structures have not been investigated in the gas phase.

the observed shortening of the SN bonds. According to the X-ray data, all the SN bonds in the S_4N_4 molecule are equivalent, with the bond distance similar to that in thiadiazole (1.62 Å)[119]. This result is in harmony with the predictions of semi-empirical CNDO/2 calculations[120].

In most molecules studied, the SN double bonds prove to be 0.08 to 0.15 Å shorter than the formally single bonds of the thiadiazole molecule. Electron-withdrawing substituents (F, Cl) cause shortening or elongation of these bonds, depending on the atom (S or N) to which they are bonded.

In the $ClN{=}SF_2$ molecule, the F atoms bonded to S cause a shortening of the NS bond so that it becomes shorter than in the ClN=S=O molecule. A similar shortening occurs on going from $HN{=}SOMe_2$ to $ClN{=}SOF_2$. Conversely, an increase in the electron-withdrawing power of the substituents at N on going from HN=S=O to ClN=S=O results in about a 0.05 Å elongation of the NS bond. In the FN=S molecule, the NS bond distance increases to nearly the same value as that characteristic of the single bond. Note that the N=O bond in FN=O (1.136 Å[121]) is 0.04 Å shorter than that in nitrous acid (1.177 Å[122]).

The X–N=S valence angle varies within rather narrow limits (114 to 122°) in the molecules we are discussing. It is some 15° less in HNSO than the angle observed in HNCO and HNCS (Table 3).

The large values of the N=S=N and N=S=O valence angles in $Me_2S({=}NH)_2$ and $Me_2S({=}O){=}NH$ molecules, of about 135°, are noteworthy.

The nonlinear character of the N=S=X unit gives rise to a variety of conformations with respect to the N=S bond in these molecules (Fig. 5).

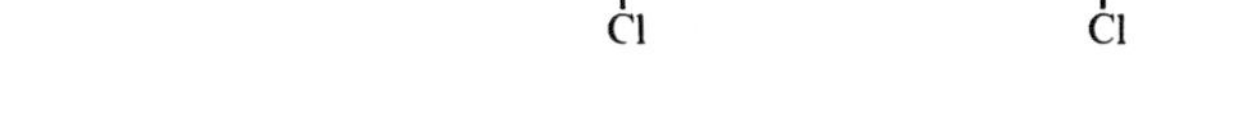

Fig. 5. Projections along a S=N bond for ClNSO, $ClNSF_2$, and $ClNSOF_2$

4. Molecules Containing ClN Bonds

Chlorine is known to occur in seven formally positive valency states. Chlorine in its higher degrees of oxidation is found primarily in chlorine fluorides or oxides and in oxygen-containing acids. As for ClN bonds, only Cl(I) derivatives have been studied.

4.1. Three-Coordinate N Atom

Average valence angles at N, α_N, in these molecules are smaller than in methylamines (Table 11), never exceed the tetrahedral angle value, and vary within rather narrow limits of 104 to 108° (Table 12). As a rule, smaller XNY angles occur in molecules containing stronger electron-withdrawing substituents. It is noteworthy that the same value of α_N was observed in both NCl_3 and NH_3 molecules. The NCl bond distances are always greater than the sum of the covalent radii, 1.72 Å, which in this case does not differ from the Schomaker-Stevenson estimate, since both atoms are characterized by the same electronegativity. Changing the number of chlorine atoms bonded to N slightly influences the NCl bond length. What is more, it remains constant to within experimental error (1.75–1.76 Å) in all the compounds studied except F_2NCl. In the latter molecule the NCl bond length decreases to 1.73 Å owing to the presence of two fluorine substituents.

Table 11. Valence angles at N atom in methylamines and ammonia

Molecule	Angle CNC, degrees	Angle HNC, degrees	Method of determination	References
H_2NMe	–	112.0(10)	MW	123)
$HNMe_2$	112.2(0.2)	108.9(0.3)	MW	123)
$HNMe_2$	111.8(0.6)	107(2)	ED	34)
NMe_3	110.6(0.6)	–	ED	34)
NH_3	–	108.2(1.1) (HNH)	ED	125)

Table 12. Structural data for molecules containing Cl–N< bonds

Molecules; symmetry of frame	ClN bond length, in Å; α_N, in degrees	Bond angles of N atom in degrees	Method	References
NCl_3 (C_{3v})	1.759(2) 107.1	ClNCl 107.1(0.5)	ED	126)
$MeNCl_2$ (C_s)	1.74(2) 108.7	ClNCl 108(2) ClNC 109(2)	ED	127)
Me_2NCl (C_s)	1.77(2) 107.3	ClNC 107(2) CNC 108(ass)	ED	127)
H_2NCl (C_s)	1.7522(1) 104.8	ClNH 103.7(0.35) HNH 107(2)	MW	128)
F_2NCl (C_s)	1.730(8) 104.3	ClNF 105(1) FNF 103(1)	ED	129)

4.2. Two-Coordinate N Atom

The influence of chlorine on the ClN=X angle in chloroisocyanate, ClN=C=O, and chloroazide ClN=N≡N (Table 13) represents a problem of great interest.

Table 13. Structural data for molecules containing Cl–N= bonds

Molecules; symmetry of frame	ClN bond length, in Å	Bond angle of N atom, in degrees	Method	References
ClN=C=O (C_s)	1.700(3)	118.2(0.6)	ED	130)
ClN=C=O (C_s)	1.703(11)	119.35(1.0)	MW	131)
ClN=N≡N (C_s)	1.745(5)	108.7(0.5) (Cl–N=N)	MW	132)
ClN=O (C_s)	1.975(5)	113.3(0.7)	MW	133)
ClN=S=O (C_1)	1.696(3)	116.3(0.4)	ED	112)
ClN=SF_2 (C_s)	1.723(4)	120.0(0.2)	ED	113)
ClN=SOF_2 (C_s)	1.713(5)	115.0(0.7)	ED	114)

It has been shown, by both microwave and electron-diffraction techniques, that these molecules possess nonlinear structures with a ClN=X angle of far less than 180°. In such studies the form of the N=XY fragments was also subjected to refinement[d]. It has been shown that deviations from linear structures, if they occur, should not exceed 10°. However, even such a possibility makes the choice of molecular models quite a complicated problem. The electron diffraction data have been found to be consistent with two sets of parameters for the ClN=C=O molecule, depending on the assumption of linear or bent geometry for the N=C=O group. The authors of[130] prefer the bent model since it gives a better fit with the microwave data and the results of semi-empirical CNDO/2 calculations.

However, both models predict a ClN=C angle of less than 125°, *i.e.* smaller than the XN=C angle in MeN=C=O and the corresponding derivatives of silicon and phosphorus. A comparison of azides reveals a similar trend. In the ClN=N≡N molecule, the ClN=N angle has been found to be as small as 108.7°.

In these compounds, the NCl bond length varies over a wider range than in the case of chloroamines. In the ClN=O molecule, the NCl bond is some 0.2 Å longer than

d) The N=XY units were assumed to be linear in other structural studies of isocyanates and azides.

in chloroamines. This phenomenon can hardly be thought of as peculiar to a given compound, since a similar trend is observed in the case of fluoro- and bromonitrosyls[121,134].

5. Molecules Containing SiO Bonds

Multiple bonding is not characteristic of silicon. The following discussion will therefore be restricted to the two-coordinate oxygen derivatives.

An experimental discovery of a drastic increase in the oxygen valence angle α_O on passing from dimethyl ether Me_2O ($\alpha_O = 111.5-111.7°$[135, 136]) to disiloxane $(H_3Si)_2O$ ($\alpha_O = 144.1°$, Table 14) generated great interest among researchers. Together with the

Table 14. Structural data for molecules containing Si–O– bonds

Molecules; symmetry of frame	SiO bond length, in Å	Bond angle of O atom, in degrees	Method	References
$(SiH_3)_2O$ (C_{2v})	1.634(2)	144.1(0.8)	ED	137)
$(SiMe_3)_2O$ (C_s)	1.63(3)	130(10)	ED	138)
$Si(OSiH_3)_4$ (Staggered)	1.63	140(5)	ED	138)
$(Me_2SiO)_3$ (C_{3v} or D_{3h})	1.635(2)	131.6(0.4)	ED	139)
$(H_2SiO)_4$ (S_4)	1.628(4)	148.6(1.2)	ED	140)
$(Me_2SiO)_4$ (S_4)	1.622(3)	144.8(1.2)	ED	139)
$(Me_2SiO)_5$	1.620(2)	146.5(1.2)	ED	139)
$(Me_2SiO)_6$	1.622(2)	149.6(1.4)	ED	139)
$(SiF_3)_2O$ (C_2)	1.580(25)	155.7(2.0)	ED	141)
$(SiCl_3)_2O$ (C_2)	1.592(10)	146(4)	ED	41)
SiH_3OMe (C_s)	1.640(3)	120.6(0.9)	ED	142)
SiH_3OPh (C_1)	1.648(7)	121(1)	ED	143)
$Si(OMe)_4$ (staggered)	1.64(3)	113(2)	ED	144)
SiF_3OMe (C_s, staggered)	1.580(fixed)	131.4(3.2)	ED	41)

increase of the α_N value on going from NMe_3 to $N(SiH_3)_3$, this fact has been treated as favoring the idea of $(p_\pi-d_\pi)$ bonding. Let us now consider the factors that govern changes in α_O in more detail. The first point to mention is that the SiOC angle is unchanged on going from H_3SiOMe to H_3SiOPh (121°), despite the fact that the former contains methyl and the latter phenyl substituents. In these molecules the angle is some 20° less than in the $(H_3Si)_2O$ molecule. In contrast to the MeOPh molecule, which is characterized by a planar skeleton[145], the H_3SiOPh molecule contains the SiH_3 group which is displaced from the Ph plane by 68(3)°, so that the assumption of an extended π-electron system must be ruled out.

Fluorination of the silyl group results in an increase of the SiOC angle up to 131° on going to F_3SiOMe. In $(F_3Si)_2O$, the effect of two fluorosilyl groups induces a further increase in the valence angle at O to 156°. However, this is only an effective angle value. There are reasons why large-amplitude, low-frequency bending vibrations can be expected in siloxanes. The equilibrium angle values may be assumed to be larger than effective ones in this and other acyclic structures discussed in this section, as was shown above to be the case with silylisocyanates. The authors of the study of the $(F_3Si)_2O$ molecule[141] think it possible that the minimum of the potential function corresponds to the linear configuration. Shrinkage correction might yield the equilibrium angle value. However, a knowledge of the bending vibration potential function is required for this purpose.

In cyclic siloxanes, the SiOSi angle is determined by the size of the ring system. The angle decreases to 131.6° in the trimeric $(OSiMe_2)_3$ molecule which is nearly planar (the OOO/OSiO dihedral angle is only 8°). The authors of the original study[139] treat this molecule as a planar one and attribute the departure from D_{3h} symmetry to out-of-plane deformations. One may wonder whether the potential surface of the molecule has two minima corresponding to inversion of equilibrium C_{3v} structures or one flat minimum with D_{3h} symmetry of equilibrium configuration. Nonplanar tetrameric species $(OSiH_2)_4$ and $(OSiMe_2)_4$ have an average SiOSi angle of about 147° which remains the same on going to the penta- and hexameric molecules $(OSiMe_2)_5$ and $(OSiMe_2)_6$. Attempts to determine the precise symmetry of the two latter molecules failed owing to large-amplitude, low-frequency vibrations of the skeleton.

The decrease in the SiOSi angle in trimeric $(OSiMe_2)_3$ from its optimum value of about 145°, as well as the nearly planar geometry of the cycle, probably are related to the strained character of the ring. Similar effects operate in the case of phosphazenes (see above).

The SiO bond distances (like the SiN ones) are considerably less than the Schomaker-Stevenson estimate for a single bond (1.76 Å).

A marked shortening (by about 0.05 Å) of the SiO bond, together with an increase in the SiOSi angle, occurs in perchloro- and perfluorodisiloxanes. A decrease of the SiOSi angle in the trimeric ring siloxane (see above) is associated with some elongation of the SiO bond as measured in other cyclic siloxanes. The discovery of the effect of valence angle widening in siloxanes aroused interest in germoxanes and stannoxanes, which have also been found to contain large valence angles at the oxygen (Table 15). However, the GeO and SnO bond distances differ considerably less from the Schomaker-Stevenson estimates than do the SiO bonds in siloxanes. The fact that the α_O value is less in $(H_3Ge)_2O$ than in $(Me_3Ge)_2O$ raises the question as to the role of steric inter-

Table 15. Electron diffraction data on oxygen derivatives of group IV elements

Molecule	Ref.	XOX angle, degrees	r (OX), Å	The Schomaker-Stevenson estimate of the OX bond distances, Å
Me_2O	135)	111.5(1.5)	1.416(3)	1.42
$(H_3Si)_2O$	137)	144.1(0.8)	1.634(2)	1.76
$(H_3Ge)_2O$	146)	126.5(0.4)	1.766(4)	1.81
$(Me_3Ge)_2O$	147)	141.0(0.5)	1.770(12)	
$(Me_3Sn)_2O$	147)	141.0(0.5)	1.940(13)	1.98

actions between Me substituents in determining the angle. Unfortunately, we cannot collate these data with the results of structure determinations of the corresponding siloxanes, since the electron-diffraction studies of the $(Me_3Si)_2O$ molecule [and also $Si(OSiH_3)_4$ and $Si(OMe)_4$] were carried out with the low-precision visual technique (see Table 14).

The symmetry of the conformations that occur in $(F_3Si)_2O$ and $(Cl_3Si)_2O$ is lowered from C_{2v} to C_2 because of opposite-sign rotations of the SiX_3 groups relative to the SiOSi plane by 34.6(1.5) and 28.9(1.5)° in fluoro and chloro derivatives, respectively. In the original paper[41)], this effect was supposed to be unrelated to the interactions of nonbonded atoms. According to the data reported, the nonbonded distances between the SiX_3 groups are greater than the sum of their van der Waals radii, the nearest F . . . F approach being 3.95(5) Å and the shortest Cl . . . Cl distance 4.01(7) Å (the values obtained from van der Waals radii are 3.0 and 3.6 Å, respectively). Considering the rather large amplitudes of the vibrations of the atom pairs X . . . X (some 0.15 to 0.20 Å) reported in this paper, one may suppose that the observed near C_2 structure corresponds to the effective configuration, while the equilibrium structure is probably C_{2v}. The observed lowering of symmetry may be related to low-frequency twisting vibrations. It appears that the potential surface of rotation about the SiO bond has a small curvature.

It is known that the threefold axes of the CH_3 and CF_3 groups are not collinear with the CO bonds in the CH_3OH[148)] and CF_3OF[149)] molecules. Attempts to detect similar effects in $(F_3Si)_2O$ and $(Cl_3Si)_2O$ were unsuccessful: the observed deviations from collinear structures amount to $0.7 \pm 5.0°$ and $3.0 \pm 3.0°$, respectively, and are insignificant compared to measuring errors. It should be noted in addition, that only slightly distorted tetrahedral configurations occur at the Si atoms in siloxanes. The effect of oxygen on the SiCl bond distance is negligible (see Table 2).

6. Molecules Containing PO Bonds

6.1. Two-Coordinate O Atom

When bonded to oxygen, phosphorus, like silicon, increases the valence angle at O, α_O.

The widest POP angle (about 140°) was observed in the $(F_2P)_2O$ molecule (Table 16). It should be noted that the two contemporary electron-diffraction studies on the $(F_2P)_2O$ molecule[150, 151] yield quite different geometric parameters. This may be attributed to difficulties inherent in interpreting the scattering patterns that arise from the exclusively strong correlations between the bond distance values, r(PF), r(PO), the root-mean-square vibrational amplitudes, l(PF), l(PO), and the POP, FPO and FPF bond angle values. On the other hand, the authors of these papers used very different approaches to interpreting the experimental data.

Table 16. Structural data for molecules containing P–O– bonds

Molecules; symmetry of frame	PO bond lengths in Å	Bond angles of O atom in degrees	Method	References
$(F_2P)_2O$ (C_1)	1.533(6)	145.1(1.2)	ED	150)
$(F_2P)_2O$ (Mixture of C_1, C_2, C_s and C_{2v})	1.631(5)	135.2(0.9)	ED	151)
$P(OMe)_3$ (C_3)	1.613(3)	119.2(1.1)	ED	152)
$P(OEt)_3$	1.600(6)	120.0(1.0)	ED	153)
$P(OCH{=}CH_2)_3$	1.600(6)	124.0(1.0)	ED	153)
$MeOPCl_2$ ($\sim C_s$)	1.63(ass.)	107(4)	ED	154)
$(MeO)_3P{=}O$ (C_3)	1.580(2)	118.4(1.6)	ED	155)
MeOP(=O)Me(F) (C_1)	1.549(7)	116(1)	ED	156)
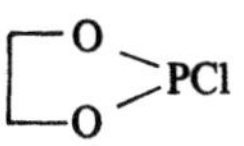 (C_s, envelope)	1.631(7)	110.6	ED	157)
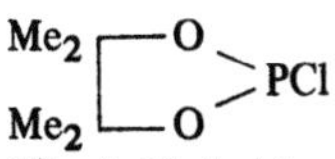 (C_1, half-chair)	1.630(10)	112,0	ED	158)
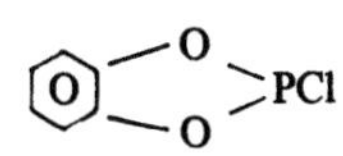 (C_s, envelope)	1.65(3)	104.6(1.0)	ED	159)

Table 16 (continued)

Molecules; symmetry of frame	PO bond lengths in Å	Bond angles of O atom in degrees	Method	References
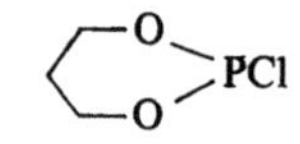 (C_s, chair)	1.63(ass.)	120.1(1.0)	ED	160)
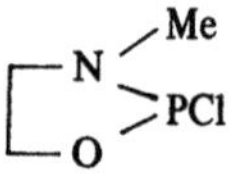 (C_1, envelope)	1.622(13)	115.6	ED	69)
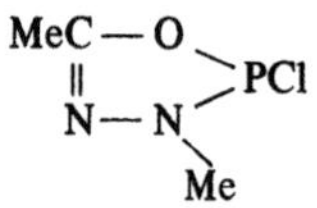 (C_1, envelope)	1.600(15)	105.5	ED	70)
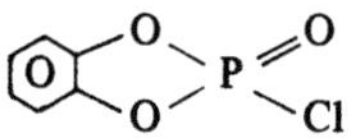 (C_s, planar cycle)	1.617(14)	107.8	ED	161)
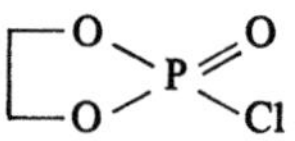 (C_1, half-chair)	1.616(10)	106.4	ED	162)
P_4O_6 (T_d)	1.638(3)	126.4(0.7)	ED	163)
$P_4O_6S_4$ (T_d)	1.61(2)	123.5(1.0)	ED	164)
P_4O_{10} (T_d)	1.604(4)	123.5(0.7)	ED	165)

In[150], the occurrence of only one conformer was suggested, and refinement of the parameters was carried out in order to arrive at the optimum conformation. The model chosen was $r(\text{PO}) < r(\text{PF}) = 1.597(4)$ Å with dihedral angles of rotation of the two PF_2 groups about the PO bonds, $\varphi_1 = -61.3$ and $\varphi_2 = 123.5°$. Fig. 6 shows the conformation of this molecule, in which $\varphi_1 = \varphi_2 = 0°$. Large-amplitude vibrations of the

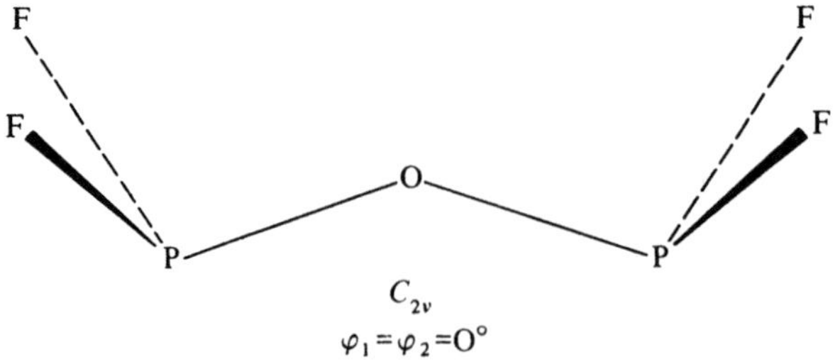

Fig. 6. Molecular model of $(F_2P)_2O$ for $\varphi_1 = \varphi_2 = 0$ (C_{2v})

atom pairs F . . . F (0.18 to 0.24 Å) were introduced to make the calculated results consistent with the experimental data. The molecular model with $r(\text{PO}) = 1.597(9) >$

r(PF) = 1.570(5) Å, ∠ POP = 137.5°, φ_1 = 56.2° and φ_2 = 122.0° was rejected.

In paper[151], the occurrence of more than one conformer was suggested by certain details of the experimental radial distribution curve. This study yielded the principal geometrical parameters of r(PO) = 1.631(5) > r(PF) = 1.568(2) Å, that is, nearly the values rejected by the authors of[150]. The final model involved a certain mixture of four conformers of C_1, C_2, C_s and C_{2v} symmetries. Lower amplitudes than those reported in[150] were obtained for F . . . F pair vibrations, *i.e.* about 0.145 Å.

Unfortunately, a choice between these two models cannot be rigorously substantiated. The condition of r(PO) > r(PF) can be regarded as indirect evidence in favor of the second model. The inverse ordering, cited in[150], that is r(PO) < r(PF) = 1.597(4) Å, has never been previously reported. Moreover, the experimental data of[151] cover a greater range of scattering angles, which makes the results of this investigation more reliable[e].

The POP angles from[150] and [151] yield an average value of 140° for the PF_2OPF_2 molecule, which is less than the SiOSi angle in $(SiF_3)_2O$, 156°. Polycyclic molecules of phosphorus oxides, which can be regarded as systems of fused 6-membered cycles, yield POP angles of about 124–126°.

The POC angle varies over a rather wide range from 105 to 125°. In the case of 5-membered cyclic phosphites and phosphates, the angle was cited to be 105 to 112° (the discrepancy in the cited values, we believe, is due mainly to the low precision of angle determination in complex ring molecules). The POC angle increases to 120° in the 6-membered cyclophosphite $H_6C_3O_2PCl$, probably because of the less strained character of the structure. Studies of $P(OMe)_3$, $P(OEt)_3$ and $(MeO)_3P{=}O$ yielded POC angle values of 118 to 120°; such values are very characteristic of acyclic compounds apparently. It is noteworthy that the higher phosphorus valency in $(MeO)_3P{=}O$ results in a certain narrowing of the POC angle on going from $P(OMe)_3$ and $P(OEt)_3$ to this compound. A similar trend (a lowering of α_N values) has been observed in the series of phosphorus amino derivatives. The largest POC angle of 124° was reported for the $P(OCH{=}CH_2)_3$ molecule. The value cited for the Cl_2POMe molecule is questionable (107 ± 4°).

The PO bond distances, like the PN distances, are considerably lower than the Schomaker-Stevenson estimate of 1.71 Å. This distance varies within about 0.05 Å in the compounds we are discussing. However, it should be remembered that no definite conclusion can yet be drawn concerning the PO bond distance in $(F_2P)_2O$ (see above). The shortening of this bond to 1.533 Å reported in[150] seems unrealistic. On the other hand, it is not quite clear why this bond remains unaffected by the adjacent PF bonds, as is the case according to[151] [r(PO) = 1.631 Å]. For instance, the SiO bond in $(F_3Si)_2O$ is considerably shorter than that in $(H_3Si)_2O$.

The increase of the PO bond lengths on going from four-coordinate to three-coordinate phosphorus derivatives is very characteristic of phosphorus compounds. Interestingly enough, longer PO bonds occur with larger POP or POC angles in this case.

---

e) The following microwave data on the $MeOPF_2$ molecule[124], published when this paper was already in print, are of interest in this connection: r (PO) = 1.560(20), r (PF) = 1.591(6) Å; ∠ POC = 123.7(0.5)°, which yields the same correlation of r (PO) and r (PF) as cited in paper[150].

Table 17. PCl bond distances in acyl chlorides of phosphorus, phosphonic and phosphoric acids

Molecule	r (PCl), Å	Method	Reference
PCl_3	2.042(2)	ED	[82]
$MeOPCl_2$	2.084(5)	ED	[154]
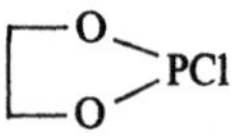	2.105(7)	ED	[157]
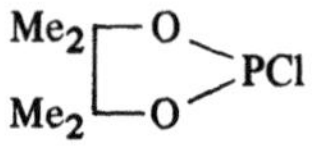	2.093(9)	ED	[158]
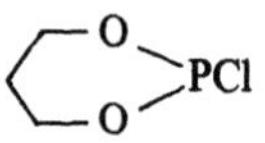	2.128(8)	ED	[160]
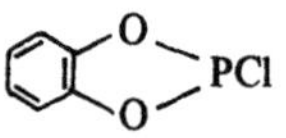	2.106(8)	ED	[159]
$Cl_3P{=}O$	1.993(3)	ED	[86]
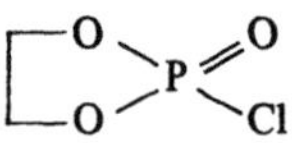	2.057(10)	ED	[162]
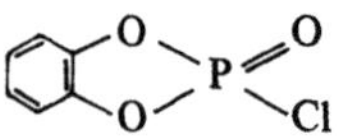	2.036(14)	ED	[161]
$MeP(=O)Cl_2$	2.032(9)	ED	[166]
$PhP(=O)Cl_2$	2.025(5)	ED	[167]
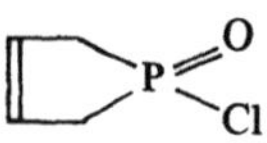	2.040(8)	ED	[168]
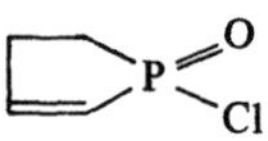	2.057(8)	ED	[169]
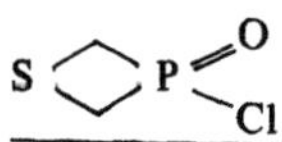	2.032(15)	ED	[170]

In addition we should note that cyclic structures contain longer PO bonds than acyclic species.

Chlorine substituents at the phosphorus atom cause relatively small changes of the PO bond lengths. Likewise small is the corresponding elongation of the PCl bonds as compared to the elongation which occurs in amino derivatives (Table 17). Changes in the number of chlorine atoms bonded to phosphorus and in the phosphorus valency state influence the PCl bond length more markedly.

The problem of rotational isomerism in the $P(OEt)_3$ and $P(OCH{=}CH_2)_3$ molecules remains unsolved owing to the highly complicated character of these molecules. In particular, rotations about two bonds (P–O and C–O) are possible in each of the ether groups involved. The $P(OMe)_3$ molecule seems to possess C_3 rather than C_{3v} symmetry; according to[152], one should consider a conformational equilibrium involving several rotational isomers for this compound. In the case of $MeOPCl_2$, the molecular symmetry appears to be lower than C_s (Fig. 7).

Phosphate molecules possess structures with a nearly trans arrangement of the O=P and OC bonds relative to the P–O bond. Thus, the OC bond is displaced from trans

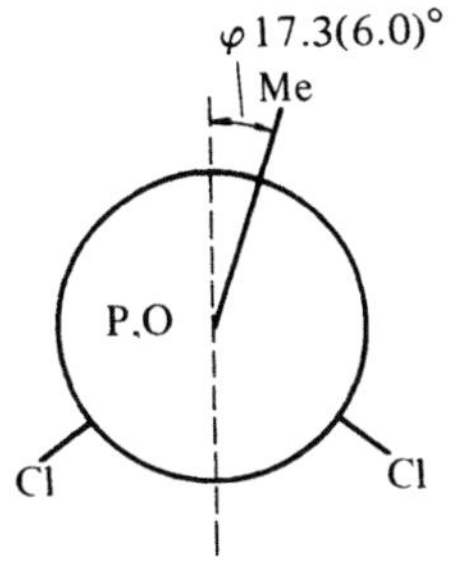

Fig. 7. Projection along a PO bond for $MeOPCl_2$

orientation by about 15° in MeOP(=O)Me(F) and 27.0° (11.4°) in the predominant (ca. 74%) isomer of $(MeO)_3P$=O against 111.8° (22.3°) in the second isomer of the latter compound.

The envelope conformation is characteristic of 5-membered cycles containing three-coordinate phosphorus (Fig. 8). In this conformation, the phosphorus atom is displaced from the plane of the remaining four ring atoms, and the P–Cl bond adopts axial orientation. However, the replacement of four H atoms of ethylenechlorophosphine by four methyl groups results in a half-chair conformation, which provides an interesting example of controlling the cycle conformation.

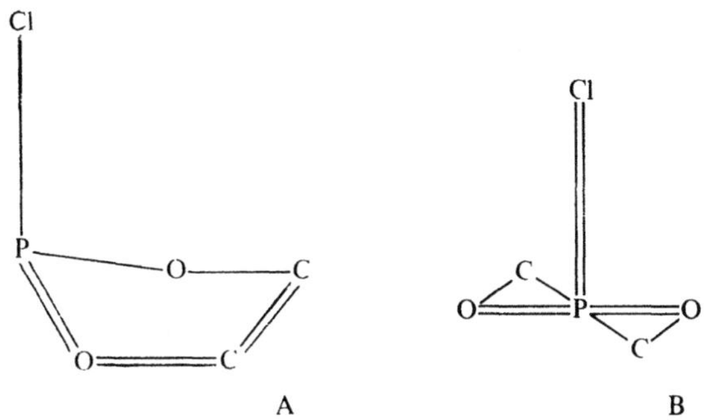

Fig. 8. "Envelope" (a) and "half-chair" (b) conformations of cycle in 2-chloro-1,3,2-dioxaphospholan

6.2. Uni-Coordinate O Atom

Multiple oxygen-phosphorus bonds are characteristic only of four-coordinate phosphorus. These bonds are considerably shorter (by about 0.15 Å) than the single PO bonds. The nature of double P=O bonding is widely discussed in terms of the more general problem of chemical bonding involving four-coordinate phosphorus[4–6].

As a rule, the P=O bond distance varies within the limits of 1.44 to 1.48 Å (Table 18). These variations provide striking examples of the influence of the chemical nature of the substituents. In the absence of electron-withdrawing substituents the P=O bond distance approaches the value of 1.48 Å, as in Me_3P=O and $(MeO)_3P$=O. In F_3P=O and F_2HP=O

Table 18. P=O bond lengths

Molecules, symmetry of frame	P=O bond length, in Å	Method	Refs.
$Me_3P=O$ (C_{3v})	1.479	ED	171)
$(MeO)_3P=O$ (C_3)	1.477(6)	ED	155)
MeOP(=O)Me(F) (C_1)	1.488(9)	ED	156)
$F_2HP=O$ (C_s)	1.437(6)	MW	172)
$F_3P=O$ (C_{3v})	1.436(6)	ED	86)
$Cl_3P=O$ (C_{3v})	1.455(5)	MW	85)
$Cl_3P=O$ (C_{3v})	1.449(5)	ED	86)
$Cl_2P(=O)N=C=O$ (C_s)	1.455(10)	ED	87)
$Me_2NP(=O)Cl_2$ (C_1)	1.47(2)	ED	63)
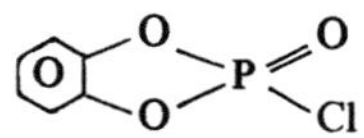 (C_s)	1.487(16)	ED	161)
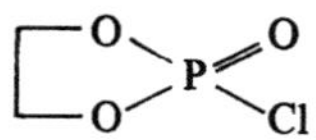 (C_1)	1.438(15)	ED	162)
$MeP(=O)Cl_2$ (C_s)	1.448(5)	ED	166)
$PhP(=O)Cl_2$ (C_s)	1.47(1)	ED	167)
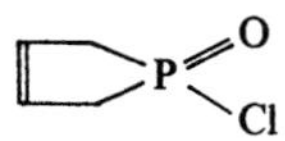 (C_s, envelope)	1.440(10)	ED	168)
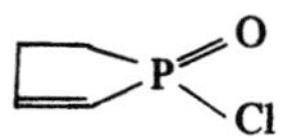 (C_1, envelope)	1.451(9)	ED	169)
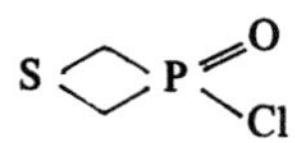 (C_s)	1.441(10)	ED	170)
P_4O_{10} (T_d)	1.429(4)	ED	165)
PO	1.4343	V-R. Sp.	173)

it becomes still shorter than 1.44 Å. The corresponding Cl derivatives contain P=O bonds ranging from 1.44 to 1.46 Å, with the exception of few results obtained with a low-precision technique.

However, the P=O bond distance in P_4O_{10}, 1.429 Å, is still shorter than in fluoro derivatives. This finding is in harmony with the data on other phosphorus oxides[163].

7. Molecules Containing SO Bonds

7.1. Two-Coordinate O Atom

Unfortunately, the available data on acyclic sulfur derivatives containing two-coordinate oxygen are limited to the four examples cited in Table 19. The sulfur atom in its highest oxidation state is bonded to the electron-withdrawing F atom in the $(SO_2F)_2O$ and $(SO_2F)_2SO_4$ molecules. The SOS angle observed in these molecules, 123.5°, is probably the largest angle that can occur at S (VI). It is noteworthy that the same value has been reported for the $S_2O_7^{2-}$, $S_3O_{10}^{2-}$, $S_5O_{16}^{2-}$ ions[180–182]. The SOF and SOO angles decrease to 108 and 105°, respectively, on passing to F_5SOF and $F_5SO{-}OSF_5$. The structures determined so far indicate that the SO bond distance varies within the range 1.61 to 1.66 Å. These bonds are considerably shorter than the Schomaker-Stevenson estimate of 1.69 Å.

Table 19. Structural data for molecules containing S–O– bonds

Molecules, symmetry of frame	SO bond lengths, in Å	Bond angle of O atom, in degrees	Method	References
$(SO_2F)_2O$ (C_1)	1.611(5)	123.6(0.5)	ED	174)
$(SO_2F)_2SO_4$ (C_2)	1.613(6)	123.6(1.2)	ED	174)
F_5SOF	1.64	108	ED	175)
$(F_5SO)_2$	1.66(5)	105(3)	ED	176)
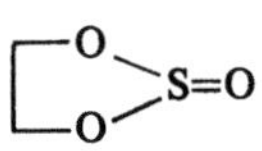 (C_s, planar cycle)	1.629(10)	108.8	ED	177)
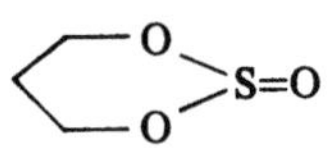 (C_s, chair)	1.62(1)	113(2)	ED	178)
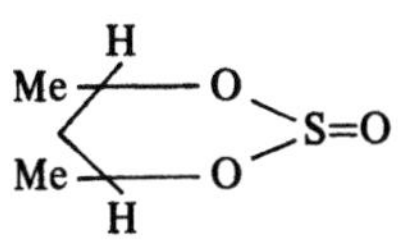 (C_s)	1.622(9)	114.1(1.0)	ED	179)

Compounds $(SO_2F)_2O$ and $(SO_2F)_2SO_4$ have been shown to possess a staggered bond configuration relative to the SO bond. In studying the structure of the former derivative, the best fit was obtained with an asymmetric model, suggesting different values for the dihedral angles FSOS (Fig. 9). The fit was less good for the model of C_2 symmetry. Lowering symmetry from C_{2v} to C_2 occurs also in $(SO_2F)_2SO_4$ (Fig. 10).

In contrast to the envelope conformation of the ethylenechlorophosphine molecule $(CH_2)_2PO_2Cl$, ethylenesulfite $(CH_2)_2SO_3$ has been shown to contain an essentially planar 5-membered cycle. The 6-membered cycle in trimethylenesulfite $(CH_2)_3SO_3$ possesses the chair conformation with the S=O bond in the axial position.

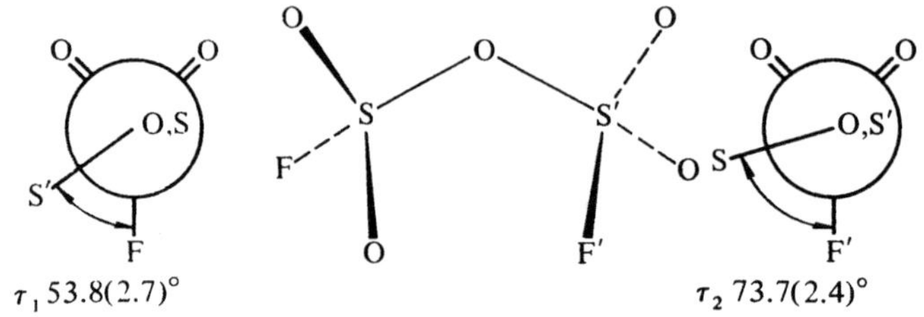

Fig. 9. Projections along the SO and S′O bonds for $(SO_2F)_2O$

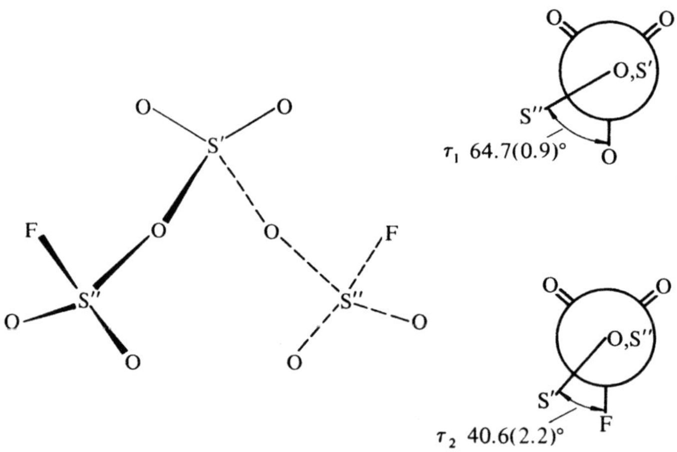

Fig. 10. Projections along the S′O and S″O bonds for $(SO_2F)_2SO_4$

7.2. Uni-Coordinate O Atom

The S=O bond distances in sulfur oxides decrease successively on passing to compounds containing sulfur in higher oxidation states, *i.e.* from 1.48 Å in the SO molecule to 1.43 and 1.42 Å in SO_2 and SO_3, respectively (Table 20). A similar decrease is observed on going from thionyl (X_2SO) to sulfuryl (X_2SO_2) derivatives. This observation bears

Molecules, symmetry of frame	S=O bond lengths, in Å	Meth.	Ref.
SO	1.4810(1)	MW	183)
SO_2 (C_{2v})	1.4349(2)	MW	184)
SO_2 (C_{2v})	1.436(1)	ED	185)
SO_2 (C_{2v})	1.431(2)	ED	186)
ClN=S=O (C_1)	1.445(4)	ED	112)
H–N=S=O (C_s)	1.451(5)	MW	111)
Me_2S=O (C_s)	1.477	MW	187)
Me_2S=O (C_s)	1.47(4)	ED	188)
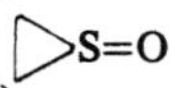 (C_s)	1.483	MW	189)
SO$(NMe_2)_2$ (C_1)	1.480(9)	ED	97)
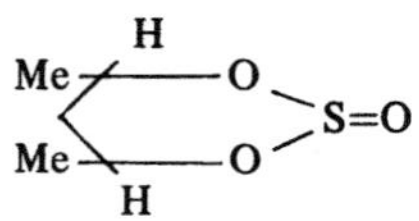 (C_s)	1.480(22)	ED	179)
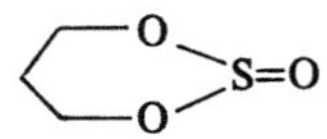 (C_s)	1.45(2)	ED	178)
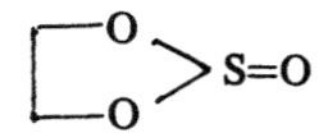 (C_s)	1.439(15)	ED	177)
SOF_2 (C_s)	1.420(3)	ED	190)
SOF_2 (C_s)	1.412(1)	MW	191)
$SOCl_2$ (C_s)	1.443(6)	ED	104)
SO_3 (D_{3h})	1.418(3)	ED	186)
Me_2SO_2 (C_{2v})	1.425(3)	ED	192)
Me_2SO_2 (C_{2v})	1.429	MW	193)
Me_2SO_2 (C_{2v})	1.431(4)	MW	194)
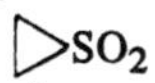 (C_{2v})	1.439	MW	195)
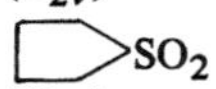 (C_{2v})	1.449(10)	ED	196)
Me_2S(=O)=NH (C_s)	1.437(3)	ED	115)
ClN=S(=O)F_2 (C_s)	1.393(4)	ED	114)
$SO_2(NMe_2)_2$ (C_{2v})	1.432(10)	ED	99)
$MeSO_2Cl$ (C_s)	1.424(3)	ED	106)
Cl_3CSO_2Cl (C_s)	1.45(1)	ED	197)
Me_2NSO_2Cl (C_s)	1.438(10)	ED	98)
SO_2Cl_2 (C_{2v})	1.404(4)	ED	104)
$MeSO_2F$ (C_s)	1.410(3)	ED	198)
SO_2FCl (C_s)	1.408(6)	MW	199)
SO_2F_2 (C_{2v})	1.405(3)	MW	200)
$(SO_2F)_2O$ (C_1)	1.398(2)	ED	174)
$(SO_2F)_2SO_4$ (C_2)	1.402(3)	ED	174)
SOF_4 (C_{2v})	1.413(8)	ED	201)
SOF_4 (C_{2v})	1.403(3)	ED	202)

relevance to a more general rule according to which thionyl compounds contain longer bonds than the corresponding sulfuryl species.

The thionyl S=O bond is more sensitive than the sulfuryl bond to the effects of other substituents at the S atom. Note that in Me_2SO and $(Me_2N)_2SO$, the S=O bond distance increases to the value observed in SO, and in Me_2SO_2 and $(Me_2N)_2SO_2$ to the value observed in SO_2.

The replacement of one or two Me or Me_2N groups by halogen atoms results in a successive decrease of the S=O bond distance, which in SOF_2 becomes less than that in SO_2, and in sulfuryl fluoride and its derivatives less than that in SO_3. However, in the case of S(VI) the S=O bond length ranges within narrower limits (1.40 to 1.45 Å) than in the case of S(IV) derivatives (1.41 to 1.48 Å). The replacement of one of the two S=O bonds in SO_2 by the S=N bond increases the remaining S=O bond by 0.01 to 0.02 Å. In the case of sulfuryl derivatives such a replacement results in a less marked elongation. Thus the S=O bond increases by only about 0.01 Å on going from Me_2SO_2 to $Me_2S(=O)=NH$. The S=O bond lengths are the same to within measuring errors in both the SO_2F_2 and $ClN=S(=O)F_2$ molecules. It should be mentioned that the S=O bond length only changes within the limits of 1.40 to 1.41 Å in S(VI) fluoroderivatives.

A marked increase in the O=S=O bond angle, associated with a shortening of the SO bonds, occurs in compounds containing the sulfuryl group on going from $(Me_2N)_2SO_2$ [114.7(2.5)°] to SO_2F_2 [124.0(0.2)°] and to $(SO_2F)_2SO_4$ [128.7(1.4)°]. The O=S=N angle amounts to 132.6(0.9)° in $Me_2S(=O)=NH$. Widening of the angle between two double bonds at S occurs simultaneously with a decrease in the angles between single bonds, from 110.5(1.3)° in $(Me_2N)_2SO_2$ to 96(0.2)° in SO_2F_2.

It should be emphasized that in some cases the electron diffraction method fails to provide reliable structural information, because the interatomic distances have nearly the same values. Thus, it has been shown[202)] that in the case of the SOF_4 molecule, four models, which differ by 10° in their FSF and FSO angles, could account for the experimental pattern. The choice between these models was made on the basis of additional considerations. These complications are probably responsible for the difference between the microwave[193, 194)] and electron-diffraction[192)] values for the OSO angle in Me_2SO_2 (121.4° and 127.1°, respectively). This problem is treated in more detail in papers on electron-diffraction studies of $MeSO_2Cl$[106)] and $MeSO_2F$[198)]. The structure of the latter compound has been resolved on the basis of the O . . . O distance from the microwave data[193)f)].

In paper[186)] electron diffraction studies of the SO_2 and SO_3 molecules have been carried out, and the equilibrium parameters of these molecules have been calculated on the basis of the experimental data. The difference between the two sets of parameters proved to be negligible (Table 21). The SO_2 and SO_3 molecules are rigid and are characterized by high frequency stretching and deformational vibrations[203)]. It is legitimate to regard effective parameters as nearly equilibrium ones, provided no low-frequency vibrations occur in the molecule. The difference between the equilibrium S=O bond

f) Now we are aware of more recent electron diffraction data on the Me_2SO_2 molecule[66)]: r (S=O) = 1.435(3), r (SC) = 1.771(4) Å; ∠ OSO = 119.(1.1)°, ∠ CSC = 102.6(0.9)°, etc. The cited paper removes the discrepancy between the microwave and electron diffraction data.

Table 21. Experimental and equilibrium parameters of SO_2 and SO_3

Parameters	SO_2			SO_3
	MW[184]	ED[186]	ED[185]	ED[186]
r_{exp}(SO), Å	1.4349(2)	1.431(2)	1.436(1)	1.418(3)
r_e(SO), Å	1.4308	1.427	1.432	1.414
$[\angle OSO]^\circ_{exp}$	119.35(0.3)	118.6(1.0)	118.4	120.2(0.6)
$[\angle OSO]^\circ_e$	119.32	118.5	118.3	120.1

lengths in SO_2 and SO_3 ($\Delta r_e = 0.017 \pm 0.003$ Å) has been predicted from quantum-mechanical calculations[204].

Now we will turn to electron diffraction studies of the metal sulfates Cs_2SO_4, K_2SO_4 and Tl_2SO_4.

The history of investigation of metal sulfates is extraordinarily interesting. The first model, suggested in an electron-diffraction study in 1965, was based on the classical structure Cs—O—SO_2—O—Cs[205]. In 1967, Buchler, Klemperer and Stauffer[206] studied Cs_2SO_4 molecular beam deflection in an inhomogeneous electric field and showed that the dipole moment of the molecule was zero to within experimental error. These authors suggested a bycyclic model with D_{2d} symmetry, with metal atoms on the opposite edges of the SO_4 tetrahedron forming two four-membered cycles (Fig. 11).

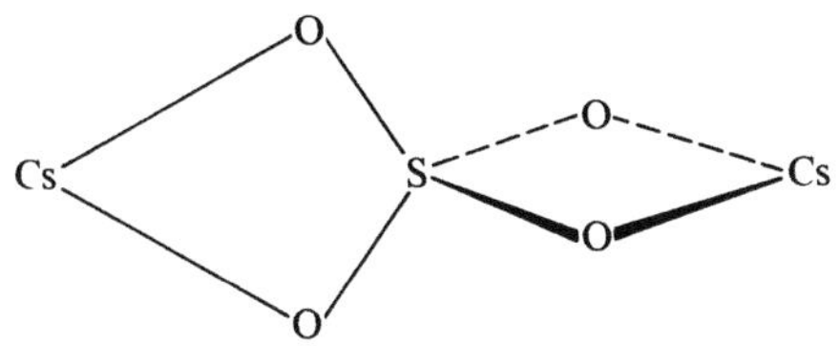

Fig. 11. Molecular model for $Cs_2SO_4(D_{2d})$

Electron-diffraction reinvestigation of Cs_2SO_4 and investigation of K_2SO_4[207] have shown that the bicyclic model is consistent with the experimental data. Assuming the SO_4 unit to be a regular tetrahedron, the authors[207] found the SO bond length to be the same as in the SO_4^{2-} ion[208], 1.49 Å. However, it proved impossible to settle the question of bond equivalency, both for the SO and CsO (KO) bonds.

The effective values of the CsO bond length (2.60 ± 0.03 Å) and KO bond length (2.45 ± 0.03 Å) in these molecules have been found to be some 0.2 Å longer than those in CsOH and KOH, respectively[209, 210]. This finding provided indirect evidence in favor of the bicyclic model, since a full set of parameters could not have been obtained for a model of D_{2d} symmetry.

In the most recent electron diffraction study on Cs_2SO_4[211], it was deduced from the large amplitude of the vibrations (l(CsO) = 0.35 Å) that the Cs_2SO_4 molecule cannot be described in terms of the so-called quasirigid model, which is characterized by effective values of bond lengths and valence angles and by amplitudes of atom pair vibrations.

A recent investigation of Tl_2SO_4[212] has yielded long interatomic distances Tl . . . O and Tl . . . Tl. However, the O . . . O distances have not been determined and the question of Tl . . . O distance equivalency has not been settled. Therefore, there are no grounds so far to consider the D_{2d} model confirmed.

8. Molecules Containing ClO Bonds

8.1. Two-Coordinate O Atom

The chlorine atom in its lowest positive oxidation state (I) causes practically no widening of the valence angle at O (Table 22). In fact, the α_O value in the Cl_2O and MeOCl molecules is the same as that in dimethyl ether Me_2O to within measuring error (111.5°[135, 136]). Rather one may suggest that chlorine (I) causes a decrease in the valence angle at O: α_O is 102.5° in the HOCl molecule against about 107° in the H_2O molecule[219]. However, α_O increases to 119° in chlorine heptaoxide, Cl_2O_7.

Table 22. Structural data for molecules containing Cl–O– bonds

Molecules, symmetry of frame	CLO bond lengths, in Å	Bond angle of O atom, in degrees	Method	References
Cl_2O (C_{2v})	1.693(3)	111.2(0.3)	ED	213)
Cl_2O (C_{2v})	1.700(1)	110.96(0.08)	MW	214)
HOCl (C_s)	1.6895(35)	102.5(0.45)	MW	215)
MeOCl (C_s)	1.674(19)	112.8(2.1)	MW	216)
Cl_2O_7 (C_2)	1.709(4)	118.6(0.7)	ED	217)
$HClO_4$ (C_{3v})	1.635(7)	113(ass.)	ED	218)

In general, the ClO bond lengths in molecules containing Cl (I) is close to the Schomaker-Stevenson estimate of 1.69 Å, and remains constant to within experimental error. However, we note that the ClO bond length decreases on going from symmetrical Cl_2O to nonsymmetrical HOCl and MeOCl species. This trend appears to be more pronounced in compounds containing Cl (VII). In the anhydride Cl_2O_7 molecule, the ClO bond is 0.07 Å longer than that in the acid $HClO_4$ molecule. Note that the length of the ClO bonds in Cl_2O and Cl_2O_7 is practically the same.

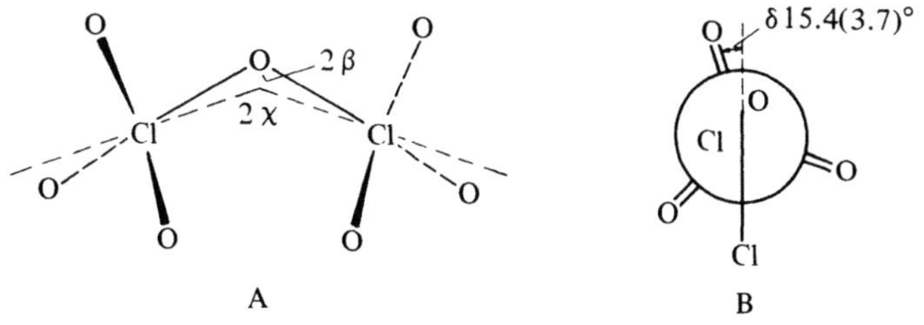

Fig. 12. The structure of Cl_2O_7. a) Molecular model of C_2 symmetry. Dotted lines are the C_3 axes of ClO_3 groups. b) Projection along the C_3 axis of a ClO_3 group on the plane that is perpendicular to the ClOCl plane.

An increase in the element-oxygen bond length on going from acids to anhydrides occurs in a number of cases. Thus, r(CO) is 1.364(3) Å in acetic acid[220] and 1.405(1) Å in the corresponding anhydride[221], which gives about 0.04 Å elongation. In the case of succinic acid, quite a similar elongation occurs[222, 223] although, unlike the acetic anhydride molecule, the succinic anhydride molecule is planar. The NO bond in the N_2O_5 molecule[224] is some 0.06 Å longer than in HNO_3[225].

The effective conformation of the Cl_2O_7 molecule shows a certain departure from C_{2v} symmetry and a deviation of the threefold axis of the ClO_3 group from the single ClO bond direction by $(\chi-\beta) = 4.7(0.8)°$ (Fig. 12). Since rather large vibrational amplitudes l(O . . . O) and l(Cl . . . O) have been determined, it is difficult to draw any conclusion as to the difference between effective and equilibrium structures.

8.2. Uni-Coordinate O Atom

The Cl=O bond length undergoes sufficient shortening when passing to compounds containing chlorine in higher oxidation states: from 1.57 Å in ClO to 1.40 or 1.41 Å in molecules containing Cl (VII) (Table 23). On the other hand, all the Cl (VII) derivatives contain Cl=O bonds of the same length.

The valence angles OClO in the Cl_2O_7 and $FClO_3$ molecules are greater than a tetrahedral angle though less than 120°, *i.e.* the ClO_3 groups represent flattened pyramids. Experimental data on the $HClO_4$ molecule proved insufficient to discriminate between the two possible models with ∠ OClO of 112.8 (0.5)° and 117.3 (0.5)°, respectively. The difficulty arises from the hardly different O . . . O and O . . . O (H) distances. Indirect considerations based on trends observed in the geometry of tetrahedral fragments XO_4 (X=Si, P, S and Cl)[231] have been utilized to arrive at a definite conclusion as to the more realistic model. The parameters of the latter are listed in Table 23. However, one cannot be very sure that the chosen model can be relied upon.

The experimental values of r(Cl=O) in the Cl (VII) derivatives are considerably less than the Schomaker-Stevenson estimate of 1.48 Å. Quite a number of authors point to a significant difference in bond lengths between the ClO and Cl=O bonds in these molecules, which amounts to about 0.3 Å. This difference disappears on going to the ClO_4^- ion.

Table 23. Cl=O bond lengths.

Molecules, symmetry of frame	Cl=O bond length, in Å	Method	References
ClO	1.569(1)	MW	226)
ClO_2 (C_{2v})	1.475(3)	ED	227)
ClO_2 (C_{2v})	1.4756	MW	228)
$FClO_2$ (C_s)	1.418(2)	MW	229)
Cl_2O_7 (C_2)	1.405(2)	ED	217)
$HClO_4$ (C_{3v})	1.407(2)	ED	218)
$FClO_3$ (C_{3v})	1.404(2)	ED	230)

Discussion

Gas-phase structural data are currently available for compounds involving most of the possible types of bonds between Si, P, S or Cl and oxygen or nitrogen. An analysis of the values of valence angles at three-coordinate nitrogen (α_N) and two-coordinate oxygen (α_O) reveals some stereochemical patterns. Thus, there appears to be a tendency for α_N and α_O values to decrease along the series of the third-row elements from Si to Cl (Tables 24 and 25).

P and S amino derivatives exemplify the decrease of the α_N value associated with an increase of the third-row element valency. This finding allows a semiquantitative prediction of an essentially pyramidal (α_N of ca. 107°) configuration of the N atom in amides of chloric acid, similar to its configuration in chloroamines.

However, in the case of chlorine oxygen derivatives, this rule does not hold, since the α_O value increases on going from Cl_2O to Cl_2O_7. Besides, there are no data on the acyclic systems of the type X–O–X, where X is P or S in oxidation state V or II and IV, respectively. Further investigation is required to reveal stereochemical patterns in more detail.

The lack of agreement between the experimental bond distances and the Schomaker-Stevenson estimates indicates that the additivity principle based on atomic increments is not very appropriate for a wide range of compounds. Shorter bond distances than the estimated ones are usually regarded as an indication of higher bond order. However, this approach fails to explain the elongation of the bond distances above the estimated

Table 24. Mean values of bond angles of N atom (α_N, degrees) and XN Single bond lengths (Å) in acyclic aminoderivatives (X = Si, P, S or Cl)

Atom X / State of oxidation	Si	P	S	Cl
I				104.3–108.7 1.73–1.77
II			117.4 1.688	–
III		115.3–120 1.68–1.70	–	[107][3])
IV	117.0–120 1.654–1.738	–	115.4 1.693	–
V		116 1.67	–	[107][3])
VI			112–116[1]) 107.0[2]) 1.65–1.69 1.697	–
VII				[107][3])

[1]) Four-coordinate S atom.
[2]) Six-coordinate S atom.
[3]) evaluation.

Table 25. Bond angles of O atom (XOX, degrees) and XO Single bond lengths (Å) in acyclic molecules (X=Si, P, S, Cl)

Atom X / State of oxidation	Si	P	S	Cl
I				111.0–111.2 1.693–1.700
II			?	–
III		135.2–145.1 1.533–1.631	–	?
IV	144.1–155.7 1.58–1.634	–	?	–
V		?	–	?
VI			123.6[1]) 1.611–1.613	–
VII				118.6 1.709

[1]) Four-coordinate S atom.

values. The most pronounced elongation has been observed with the NCl, PCl and SCl bonds. The additivity principle based on the contributions from molecular fragments, and in particular from chemical bonds, holds much better, provided the valency states of the atoms participating in the chemical bonding of a given type and the nature of the nearest environment are taken into account.

The considerable increase in the chlorine-element bond distances in the fragments >N–X–Cl (X=Si, P, S) caused by the presence of an amino group is noteworthy. A similar elongation of the metal-halogen bonds caused by trans ligands has been studied in transition-metal complexes[232–234]. One could suggest that the influence of the amino group on the chlorine-element bonds represents an intermediate case of interatomic interactions, having features characteristic of both organic and inorganic compounds.

The N atom in planar environment can be regarded as realizing a valency state (sp^2) different from that characteristic of pyramidal coordination of nitrogen (sp^3). This change of valency state is not associated with a change of coordination number, in contrast to the case of carbon $-\overset{|}{\underset{|}{C}}-$, >C– or ≥C–. In this connection it is important to note that the NC bond of the amino group does not undergo any considerable change when passing from an essentially pyramidal configuration at N to planar coordination (Table 26).

Similarly, the CO bond distance of the methoxy group remains unaffected in H_3SiOMe, despite the considerable increase in the oxygen valence angle (Table 27). It

Table 26. NC bond lengths (Å) and α_N (degrees) in amines

Molecules	r(NC)	α_N	Ref.
$HNMe_2$	1.462(5)	110.0	123)
$HNMe_2$	1.455(2)	108.6	124)
$ClSi(NMe_2)_3$	1.462(7)	119.8	33)
$(H_3Si)_2NMe$	1.465(5)	120.0	24)
H_3SiNMe_2	1.462(4)	117.0	29)
$P(NMe_2)_3$	1.458(5)	117.5	61)
$S(NMe_2)_2$	1.473(6)	117.4	96)

Table 27. OC bond lengths (Å) and α_O (degrees) in ethers

Molecules	r(OC)	α_O	Ref.
Me_2O	1.416(3)	111.5(1.5)	135)
Me_2O	1.410(3)	111.7(0.35)	136)
H_3SiOMe	1.418(9)	120.6(0.9)	142)
H_3SiOPh	1.357(9)	121.0(1.0)	143)
PhOMe	1.361(15) [OC_{Ph}]	120.0(2.0)	145)

Table 28. X=O bond lengths (Å, X=P, S, Cl)

X State of oxidation	P	S	Cl
(II)[1]	1.4343	1.4810	1.569
(IV)	–	1.412–1.483	1.475
(V)	1.429–1.488	–	1.434
(VI)	–	1.398–1.450	–
(VII)	–	–	1.404–1.408

1) Twoatomic molecules XO.

is noteworthy that the OC_{Ph} bond distance is the same in both the planar MeOPh molecule and nonplanar H_3SiOPh, where the OSi bond forms a 70° angle with the phenyl ring plane.

The X=O double bond distances (X=P, S, Cl) strongly depend on the valency state of X and on the environment (Table 28). Increasing the oxidation number of X decreases the X=O bond length.

The above discussion points to the need for more detailed studies of stereochemical patterns and emphasizes the importance of such investigations in developing concepts of the nature of chemical interactions in the molecule.

Acknowledgement. The authors are indebted to Dr. A. Haaland who suggested this investigation. We are grateful to Dr. B. Beagley, Prof. D. W. J. Cruickshank, Dr. I. Hargittai, Dr. V. A. Naumov, Dr. H. Oberhammer, Prof. W. H. Rankin and Prof. G. H. Sheldrick for reprints of their papers, which proved very useful in writing this review.

References

1) Pauling, L.: Nature of the chemical bond. Ithaca, Cornell University Press 1939.
2) Wells, A. F.: Structural inorganic chemistry. Oxford: Clarendon Press 1945.
3) Shomaker, V., Stevenson, D. P.: J. Am. Chem. Soc., *63*, 37 (1941).
4) Cruickshank, D. W. J.: J. Chem. Soc. 5486 (1961).
5) Bartell, L. S., Su, L. S., Yow, H.: Inorg. Chem. *9*, 1903 (1970).
6) Mitchell, K. A. R.: Chem. Rev. *69*, 157 (1969).
7) Ebsworth, E. A. V.: Organometallic compounds of the group IV elements, Vol. 1. New York: Marcel Dekker Inc. 1968.
8) Collins, G. A. D., Cruickshank, D. W. J., Areeze, A.: Trans. Farad. Soc. *68*, 1189 (1972).
9) Gibbs, G. V., Hamil, M. M., Louisnathan, S. J., Bartell, L. S., Yow, H.: Amer. Mineral. *57*, 1578 (1972).
10) Czismadia, I. G., Tel, L. M., Cowley, A. H., Taylor, M. W., Wolf, S.: J. C. S. Chem. Comm. 1147 (1972).

11) Oberhammer, H.: Theoret. Chim. Acta *26,* 79 (1972).
12) Guimon, C., Goubean, D., Pfister-Guillouzo, G.: J. Mol. Struct. *16,* 271 (1973).
13) Dyatkina, M. E., Klimenko, N. M.: J. Struct. Chem. (Russian) *14,* 173 (1973).
14) Charkin, O. P.: J. Struct. Chem. (Russian), *14,* 389 (1973).
15) Almenningen, A., Bastiansen, O., Traetteberg, M.: Acta Chem. Scand. *13,* 1699 (1959).
16) Morino, Y.: Acta Cryst. *13,* 1107 (1960).
17) Cyvin, S. J.: Molecular vibrations and mean square amplitudes. Oslo: Universitetsforlaget, Amsterdam: Elsevier Publishing Company 1968.
18) Hedberg, K.: J. Am. Chem. Soc. *77,* 6491 (1955).
19) Beagley, B., Conrad, A. R.: Trans. Far. Soc. *66,* 2740 (1970).
20) Rankin, D. W. H., Robiette, A. G., Sheldrick, G. M., Sheldrick, W. S., Aylett, B. J., Ellis, I. A., Monaghan, J. J.: J. Chem. Soc. *A,* 1224 (1969).
21) Robiette, A. G., Sheldrick, G. M., Sheldrick, W. S., Beagley, B., Cruickshank, D. W. J., Monaghan, J. J., Aylett, B. J., Ellis, I. A.: Chem. Comm. *1968* 909.
22) Golubinskij, A. V., Rozsondai, B., Hargittai, I., Mastryukov, V. S., Vilkov, L. V.: The abstracts of the 2-nd European Crystallografic Meeting. August 1974, Hungary.
23) Yokoi, M., Yamasaki, K.: J. Am. Chem. Soc. *75,* 4139 (1953).
24) Glidewell, C., Rankin, D. W. H., Robiette, A. G., Sheldrick, G. M., J. Mol. Struct. *4,* 215 (1969).
25) Glidewell, C., Rankin, D. W. H., Robiette, A. G., Sheldrick, G. M.: J. Chem. Soc. *A,* 318 (1970).
26) Clark, A. H., Haaland, A.: Acta Chem. Scand. *24,* 3024 (1970).
27) Robiette, A. G., Sheldrick, G. M., Sheldrick, W. S.: J. Mol. Struct. *5,* 423 (1970).
28) Roper, W. R., Wilkins, C. J.: Trans. Far. Soc. *58,* 1686 (1962).
29) Glidewell, C., Rankin, D. W. H., Robiette, A. G.; Sheldrick, G. M.: J. Mol. Struct. *6,* 231 (1970).
30) Glidewell, C., Robiette, A. G., Sheldrick, G. M.: J. Mol. Struct. *9,* 476 (1971).
31) Arnold, D. E., Ebsworth, E. A. V., Jessep, H. F., Rankin, D. W. H.: JCS Dalton Tr. 1681 (1972).
32) Airey, W., Glidewell, C., Robiette, A. G., Sheldrick, G. M., Freeman, J. M.: J. Mol. Struct. *8,* 423 (1971).
33) Vilkov, L. V., Tarasenko, N. A.: Chem. Comm. 1176 (1969).
34) Beagley, B., Hewitt, T. G.: Trans. Far. Soc. *64,* 2561 (1968).
35) Sujishi, S., Witz, S.: J. Am. Chem. Soc. *76,* 4631 (1954).
36) Glidewell, C., Rankin, D. W. H., Robiette, A. G.: J. Chem. Soc. *A,* 2935 (1970).
37) Beagley, B., Robiette, A. G., Sheldrick, G. M.: J. Chem. Soc. *A,* 3002, 3006 (1968).
38) Townes, C. N., Schawlow, A. L.: Microwave spectroscopy. New York: McGraw-Hill 1955.
39) Morino, Y., Murata, Y.: Bull. Chem. Soc. Japan *38,* 104 (1965).
40) Mockler, R. C., Batley, J. H., Gordy, W.: J. Chem. Phys. *21,* 1710 (1953).
41) Airey, W., Glidewell, C., Robiette, A. G., Sheldrick, G. M.: J. Mol. Struct. *8,* 413 (1971).
42) Bak, B., Bruhn, J., Rastrup-Andersen, J.: Acta Chem. Scand. *8,* 367 (1954).
43) Sheridan, J., Gordy, W.: J. Chem. Phys. *19,* 965 (1951).
44) Hilderbrandt, R. L., Bauer, S. H.: J. Mol. Struct. *3,* 325 (1969).
45) Jones, L. H., Schoolery, J. W., Schulman, R. G., Yost, D. M.: J. Chem. Phys. *18,* 990 (1950).
46) Kewley, R., Sastry, K. V. L. N., Winnewisser, M.: J. Mol. Spectrosc. *10,* 418 (1963).
47) Winnewisser, M., Cook, R. L.: J. Chem. Phys. *41,* 999 (1964).
48) Anderson, D. W. W., Rankin, D. W. H., Robertson, A.: J. Mol. Struct. *14,* 385 (1972).
49) Gerry, M. C., Thompson, J. C., Sugden, T. M.: Nature *211,* 846 (1966).
50) Glidewell, C., Robiette, A. G., Sheldrick, G. M.: Chem. Phys. Letters *16,* 526 (1972).
51) Jenkins, D. R., Kewley, R., Sugden, T. M.: Trans. Far. Soc. *58,* 1284 (1962).
52) Ebsworth, E. A. V., Mays, M. J.: J. Chem. Soc. 3450 (1964).
53) Glidewell, C., Robiette, A. G., Sheldrick, G. M.: Personal communication.
54) Kimura, K., Katada, K., Bauer, S. H.: J. Am. Chem. Soc. *88,* 416 (1966).
55) Dakkouri, M., Oberhammer, H.: Z. Naturforsch. *27A,* 1331 (1972).
56) Airey, W., Glidewell, C., Robiette, A. G., Sheldrick, G. M.: J. Mol. Struct. *8,* 435 (1971).

57) Hjoritas, K.: Acta Chem. Scand. *21*, 1381 (1967).
58) Murdoch, J. D., Rankin, D. W. H.: J. C. S. Chem. Comm. 748 (1972).
59) Khaikin, L. S., Vilkov, L. V.: Chem. Rev. (Russian) *40*, 2174 (1971).
60) Khaikin, L. S., Vilkov, L. V.: Chem. Rev. (Russian) *41*, 2224 (1972).
61) Vilkov, L. V., Khaikin, L. S., Evdokimov, V. V.: J. Struct. Chem. (Russian) *13*, 7 (1972).
62) Zaripov, N. M., Naumov, V. A., Tuzova, L. L.: Phosphorus. *4*, 179 (1974).
63) Vilkov, L. V., Khaikin, L. S.: Doklady Acad. Sci. USSR *168*, 810 (1966).
64) Brittain, A. H., Smith, J. E., Lee, P. L., Cohn, K., Schwendeman, R. H.: J. Am. Chem. Soc. *93*, 6772 (1971).
65) Holywell, G. C., Rankin, D. W. H., Beagley, B., Freeman, J. M.: J. Chem. Soc. *A*, 785 (1971).
66) Hargittai, M., Hargittai, I.: J. Mol. Struct. *20*, 283 (1974).
67) Forti, P., Damiani, D., Favero, P. G.: J. Am. Chem. Soc. *95*, 756 (1973).
68) Naumov, V. A., Gulyaeva, N. A., Pudovick, M. A.: Doklady Acad. Sci. USSR *203*, 590 (1972).
69) Naumov, V. A., Pudovick, M. A.: Doklady Acad. Sci. USSR *203*, 351 (1972).
70) Vilkov, L. V., Khaikin, L. S., Vasil'ev, A. F., Italinskaya, T. L., Mel'nikov, N. M., Negrebetsky, V. V., Shvetsov-Shilovsky, N. I.: Doklady Acad. Sci. USSR *187*, 1293 (1969).
71) Vilkov, L. V., Khaikin, L. S., Vasil'ev, A. F., Ignatova, N. P., Mel'nikov, N. N., Negrebetsky, V. V., Shvetsov-Shilovsky, N. I.: Doklady Acad. Sci. USSR *197*, 1081 (1971).
72) Vasil'ev, A. F., Vilkov, L. V., Ignatova, N. P., Mel'nikov, N. N., Negrebetsky, V. V., Shvetsov-Shilovsky, N. I., Khaikin, L. S.: Doklady Acad. Sci. USSR *183*, 95 (1968).
73) Almenningen, A., Andersen, B., Astrup, E. E.: Acta Chem. Scand. *23*, 2179 (1969).
74) Dimroth, K., Hoffmann, P.: Chem. Ber. *99*, 1325 (1966).
75) Allmann, R.: Chem. Ber. *99*, 1332 (1966).
76) Kawada, I., Allmann, R.: Angew. Chem. *80*, 40 (1968).
77) Bart, J. C. J., Daly, J. J.: J. Chem. Soc. *A*, 567 (1970).
78) Fischer, W., Hellner, E., Chatzidakis, A., Dimroth, K.: Tetrahedron Letters *1968*, 6227.
79) Kuczkowski, R. L., Ashe, A. J.: J. Mol. Spectrosc. *42*, 457 (1972).
80) Muetterties, E. L., Mahler, W., Schmutzler, R.: Inorg. Chem. *2*, 613 (1963).
81) Muetterties, E. L., Mahler, W., Packer, K. J., Schmutzler, R.: Inorg. Chem. *3*, 1298 (1964).
82) Kisliuk, P., Townes, C. H.: J. Chem. Phys. *18*, 1109 (1950).
83) Hedberg, K., Iwasaki, M.: J. Chem. Phys. *36*, 589 (1962).
84) Brittain, A. H., Smith, J. E., Schwendeman, R. H.: Inorg. Chem. *11*, 39 (1972).
85) Li, Y. S., Chen, M. M., Durig, J. R.: J. Mol. Struct. *14*, 261 (1972).
86) Moritani, T., Kuchitsu, K., Morino, Y.: Inorg. Chem. *10*, 344 (1971).
87) Naumov, V. A., Semashko, V. N., Shatrukov, L. F.: Doklady Acad. Sci. USSR *209*, 118 (1973).
88) Davis, M. I., Paul, J. W.: J. Mol. Struct. *12*, 249 (1972).
89) Durig, J. R., Casper, J. M.: J. Phys. Chem. *75*, 3837 (1971).
90) Szasz, G. J.: J. Chem. Phys. *23*, 2449 (1955).
91) Planje, M. C., Toneman, L. G., Dallinga, G.: Rec. Trav. Chim. *84*, 232 (1965).
92) Rankin, D. W. H., Cyvin, S. J.: J. C. S Dalton Tr. *1972*, 1277.
93) Rankin, D. W. H.: J. C. S. Dalton Tr. *1972*, 869.
94) Davis, M. I., Paul, J. W.: J. Mol. Struct. *9*, 476 (1971).
95) Paul, J. W.: Ph. D. Dissertation. Austin: University of Texas 1969.
96) Hargittai, I., Hargittai, M.: Acta Chim. Acad. Sci. Hungaricae *75*, 129 (1973).
97) Hargittai, I., Vilkov, L. V.: Acta Chim. Acad. Sci. Hungaricae *63*, 143 (1970).
98) Vilkov, L. V., Hargittai, I.: Acta Chim. Acad. Sci. Hungaricae *52*, 423 (1967).
99) Hargittai, I., Vajda, E., Szoke, A.: J. Mol. Struct. *18*, 381 (1973).
100) Haase, J., Oberhammer, H., Zeil, W.: Z. Naturforsch. *26A*, 1333 (1971).
101) Jordan, T., Smith, H. W., Lohr, L. L., Lipscomb, W. N.: J. Am. Chem. Soc. *85*, 846 (1963).
102) Morino, Y., Murata, Y., Ito, T., Nakamura, J.: J. Phys. Soc. Japan *17*, Suppl. B II, 37 (1962).
103) Beagley, B., Eckersley, S. H., Brown, D. P., Tomlinson, D.: Trans. Far. Soc. *65*, 2300 (1969).
104) Hargittai, I.: Acta Chim. Acad. Sci. Hungaricae *60*, 231 (1969).
105) Emken, W. C., Hedberg, K.: J. Chem. Phys. *58*, 2195 (1973).

106) Hargittai, M., Hargittai, I.: J. Chem. Phys. *59*, 2513 (1973).
107) Pierce, L., Hayashi, M.: J. Chem. Phys. *35*, 479 (1961).
108) Momany, F. A., Bonham, R. A.: J. Am. Chem. Soc. *86*, 162 (1964).
109) Dobyns, S. V., Pierce, L.: J. Am. Chem. Soc. *85*, 3553 (1963).
110) Rogowski, F.: Z. Phys. Chem. (BRD) *27*, 227 (1961).
111) Kirchhoff, W. H.: J. Am. Chem. Soc. *91*, 2437 (1969).
112) Oberhammer, H.: Z. Naturforsch. *25A*, 1497 (1970).
113) Haase, J., Oberhammer, H., Zeil, W., Glemser, O., Mews, R.: Z. Naturforsch. *25A*, 153 (1970).
114) Oberhammer, H.: Private communication.
115) Oberhammer, H., Zeil, W.: Z. Naturforsch. *25A*, 845 (1970).
116) Oberhammer, H., Zeil, W.: Z. Naturforsch. *24A*, 1612 (1969).
117) Kirchhoff, W. H., Wilson, E. B.: J. Am. Chem. Soc. *85*, 1726 (1963).
118) Kirchhoff, W. H., Wilson, E. B.: J. Am. Chem. Soc. *84*, 334 (1962).
119) Clark, D.: J. Chem. Soc. *1952*, 1615.
120) Cassoux, P., Labarre, J. F.: J. Mol. Struct. *13*, 405 (1972).
121) Buckton, K. S., Legon, A. C., Millen, D. J.: Trans. Far. Soc. *65*, 1975 (1969).
122) Cox, A. P., Brittain, A. H., Finnigan, D. J.: Trans. Far. Soc. *67*, 2179 (1971).
123) Wallrab, J. E., Laurie, V. W.: J. Chem. Phys. *48*, 5058 (1968); *51*, 1580 (1969).
124) Codding, E. G., Jones, C. E., Schwendeman, R. H.: Inorg. Chem. *13*, 178 (1974).
125) Kuchitsu, K., Gillory, J. P., Bartell, L. S.: J. Chem. Phys. *49*, 2488 (1968).
126) Burgi, H. B., Stedman, D., Bartell, L. S.: J. Mol. Struct. *10*, 31 (1971).
127) Stevenson, D. P., Schomaker, V.: J. Am. Chem. Soc. *62*, 1913 (1940).
128) Cazzoli, G., Lister, D. J., Favero, P. J.: J. Mol. Spectroscopy *42*, 286 (1972).
129) Vilkov, L. V., Nazarenko, I. I.: J. Struct. Chem. (Russian) *8*, 346 (1967).
130) Oberhammer, H.: Z. Naturforsch. *26A*, 280 (1971).
131) Hocking, W. H., Gerry, M. C. L.: J. Mol. Spectroscopy *42*, 547 (1972).
132) Cook, R. L., Gerry, M. C. L.: J. Chem. Phys. *53*, 2525 (1970).
133) Millen, D. J., Pannell, J.: J. Chem. Soc. *1961*, 1322.
134) Millen, D. J., Mitra, D.: Trans. Far. Soc. *66*, 2414 (1970).
135) Kimura, K., Kubo, M.: J. Chem. Phys. *30*, 151 (1959).
136) Blukis, U., Kasai, P. H., Myers, R. J.: J. Chem. Phys. *38*, 2753 (1963).
137) Almeningen, A., Bastiansen, O., Ewing, V., Hedberg, K., Traetteberg, M.: Acta Chem. Scand. *17*, 2455 (1963).
138) Yokoi, M.: Bull. Chem. Soc. Japan *30*, 100 (1957).
139) Oberhammer, H., Zeil, W., Fogarasi, G.: J. Mol. Struct. *18*, 309 (1973).
140) Glidewell, C., Robiette, A. G., Sheldrick, G. M.: Chem. Comm., *1970*, 931.
141) Airey, W., Glidewell, C., Rankin, D. W. H., Robiette, A. G., Sheldrick, G. M.: Trans. Far. Soc. *66*, 551 (1970).
142) Glidewell, C., Rankin, D. W. H., Robiette, A. G., Sheldrick, G. M., Beagley, B., Freeman, J. M.: J. Mol. Struct. *5*, 417 (1970).
143) Glidewell, C., Rankin, D. W. H., Robiette, A. G., Sheldrick, G. M., Beagley, B., Freeman, J. M.: Trans. Far. Soc. *65*, 2621 (1969).
144) Yamasaki, K., Kotera, A., Yokoi, M., Ueda, Y.: J. Chem. Phys. *18*, 1414 (1950).
145) Seip, H. M., Seip, R.: Acta Chem. Scand. *27*, 4024 (1973).
146) Glidewell, C., Rankin, D. W. H., Robiette, A. G., Sheldrick, G. M., Beagley, B., Cradock, S.: J. Chem. Soc. *A*, 315 (1970).
147) Vilkov, L. V., Tarasenko, N. A.: J. Struct. Chem. (Russian) *10*, 1102 (1969).
148) Venkateswarlu, P., Gordy, W.: J. Chem. Phys. *23*, 1200 (1955).
149) Diodati, F. P., Bartell, L. S.: J. Mol. Struct. *8*, 383 (1971).
150) Arnold, D. E. L., Rankin, D. W. H.: J. Fluorine Chem. *2*, 405 (1972/1973).
151) Bartell, L. S.: Private communication.
152) Oberhammer, H.: Private communication.
153) Khaikin, L. S., Vilkov, L. V.: J. Struct. Chem. (Russian) *10*, 722 (1969).
154) Naumov, V. A., Zaripov, N. M., Dashevskii, V. G.: J. Struct. Chem. (Russian), *12*, 158 (1971).
155) Oberhammer, H.: Z. Naturforsch. *28A*, 1140 (1973).

156) Zeil, W., Kratz, H., Haase, J., Oberhammer, H.: Z. Naturforsch. *28A*, 1717 (1973).
157) Naumov, V. A., Zaripov, N. M., Dashevskii, V. G.: Doklady Acad. Sci. USSR *188*, 1062 (1969).
158) Naumov, V. A., Zaripov, N. M.: J. Struct. Chem. (Russian) *11*, 1108 (1970).
159) Arbuzov, B. A., Naumov, V. A., Shaidullin, S. A., Muckmenev, E. G.: Doklady Acad. Sci. USSR *204*, 859 (1972).
160) Naymov, V. A., Zaripov, N. M.: J. Struct. Chem. (Russian) *13*, 768 (1972).
161) Naumov, V. A., Shaidullin, S. A.: J. Struct. Chem. (Russian), *15*, 133 (1974).
162) Naumov, V. A., Semashko, V. N., Zav'yalov, A. P., Cherkasov, R. A., Grishina, L. N.: J. Struct. Chem. (Russian) *14*, 787 (1973).
163) Beagley, B., Cruickshank, D. W. J., Hewitt, T. G., Jost, K. H.: Trans. Far. Soc. *65*, 1219 (1969).
164) Stosick, A. J.: J. Am. Chem. Soc. *61*, 1130 (1939).
165) Beagley, B., Cruickshank, D. W. J., Hewitt, T. G., Haaland, A.: Trans. Far. Soc. *63*, 836 (1967).
166) Naumov, V. A., Semashko, V. N.: J. Struct. Chem. (Russian) *12*, 317 (1971).
167) Vilkov, L. V., Sadova, N. I., Zilberg, I. Ju.: J. Struct. Chem. (Russian) *8*, 528 (1967).
168) Naumov, V. A., Semashko, V. N.: J. Struct. Chem. (Russian) *11*, 979 (1970).
169) Naumov, V. A., Semashko, V. N.: Doklady Acad. Sci. USSR *193*, 348 (1970).
170) Naumov, V. A., Semashko, V. N.: Doklady Acad. Sci. USSR *200*, 882 (1971).
171) Wang, H. K.: Acta Chem. Scand. *19*, 879 (1965).
172) Centofanti, L. F., Kuczkowski, R. L.: Inorg. Chem. *7*, 2582 (1968).
173) Coquart, B., Prudhomme, J. C.: C. r. Acad. Sci. (France) *275B*, 383 (1972).
174) Hencher, J. L., Bauer, S. H.: Can. J. Chem. *51*, 2047 (1973).
175) Crawford, R. A., Dudley, F. B., Hedberg, K.: J. Am. Chem. Soc. *81*, 5287 (1959).
176) Harvey, R. B., Bauer, S. H.: J. Am. Chem. Soc. *76*, 859 (1954).
177) Arbuzov, B. A., Naumov, V. A., Zaripov, N. M., Pronicheva, L. D.: Doklady Acad. Sci. USSR *195*, 1333 (1970).
178) Naumov, V. A., Zaripov, N. M., Shatrukov, L. F.: J. Struct. Chem. (Russian) *11*, 579 (1970).
179) Mustoe, F. J., Hencher, J. L.: Can. J. Chem. *50*, 3892 (1972).
180) Lynton, H. R., Truter, M. R.: J. Chem. Soc. *1960*, 5112.
181) Cruickshank, D. W. J.: Acta Cryst. *17*, 684 (1964).
182) DeVries, R., Mijlkoff, F. C.: Acta Cryst. *25B*, 1696 (1969).
183) Powell, F. X., Lide, D. R.: J. Chem. Phys. *41*, 1413 (1964).
184) Morino, Y., Kukuchi, Y., Saito, S., Hirota, E.: J. Mol. Spectroscopy *13*, 95 (1964).
185) Haase, J., Winnewisser, M.: Z. Naturforsch. *23A*, 61 (1968).
186) Clark, A. H., Beagley, B.: Trans. Far. Soc. *67*, 2216 (1971).
187) Dreizler, H., Dendl, G.: Z. Naturforsch. *19A*, 512 (1964).
188) Bastiansen, O., Viervoll, H.: Acta Chem. Scand. *2*, 702 (1948).
189) Saito, S.: Bull. Chem. Soc. Japan *42*, 663 (1969).
190) Hargittai, I., Mijlhoff, F. C.: J. Mol. Struct. *16*, 69 (1973).
191) Ferguson, R. C.: J. Am. Chem. Soc. *76*, 850 (1954).
192) Oberhammer, H., Zeil, W.: J. Mol. Struct. *6*, 399 (1970).
193) Jacob, E. J., Lide, D. R.: J. Chem. Phys. *54*, 459 (1971).
194) Saito, S., Makino, F.: Bull. Chem. Soc. Japan *45*, 92 (1972).
195) Nakano, Y., Saito, S., Morino, Y., Bull. Chem. Soc. Japan *43*, 368 (1970.
196) Naumov, V. A., Semashko, V. N., Shaidullin, S. A.: J. Struct. Chem. (Russian) *14*, 595 (1973).
197) Alekseev, N. V.: J. Struct. Chem. (Russian) *8*, 532 (1967).
198) Hargittai, I., Hargittai, M.: J. Mol. Struct. *15*, 399 (1973).
199) Holt, C. W., Gerry, M. C. L.: Chem. Phys. Lett. *9*, 621 (1971).
200) Lide, D. R., Mann, D. E., Fristrom, R. M.: J. Chem. Phys. *26*, 734 (1957).
201) Hencher, J. L., Cruickshank, D. W. J., Bauer, S. H.: J. Chem. Phys. *48*, 518 (1968).
202) Gunderson, G., Hedberg, K.: J. Chem. Phys. *51*, 2500 (1969).
203) Nakamoto, K.: Infrared spectra of inorganic and coordination compounds. New York: J. Wiley and Sons, Inc. 1963,
204) Moffit, W.: Proc. Roy. Soc. *A 200*, 409 (1950).

205) Spiridonov, V. P., Khodchencov, A. H., Akishin, P. A.: J. Struct. Chem. (Russian) *6*, 633 (1965).
206) Buchler, A., Stauffer, J. L., Klemperer, W.: J. Chem. Phys. *46*, 605 (1967).
207) Spiridonov, V. P., Lutoshkin, B. I.: Vestnik Moscow University, Chem. Ser. *1970*, 509.
208) Tables of Interatomic Distances and Configuration in Molecules and Ions (Sci. ed. Sutton, L. E.). The Chemical Society, Spec. Publ. 11 and 18, London, 1958–1965.
209) Lide, D. R., Kuczkowski, R. L.: J. Chem. Phys. *46*, 4768 (1967).
210) Kuczkowski, R. L., Lide, D. R., Krisher, L. C.: J. Chem. Phys. *44*, 3131 (1966).
211) Ugarov, V. V., Tolmatchev, S. M., Ezhov, Yu. S., Rambidi, N. G.: 4th Austin Symp. on Gas Phase Mol. Struct. Austin, Texas, USA: Abstract of Paper W11, p. 78 (1972).
212) Ugarov, V. V., Ezhov, Yu. S., Rambidi, N. G.: J. Struct. Chem. (Russian) *14*, 548 (1973).
213) Beagley, B., Clark, A. K., Hewitt, T. G.: J. Chem. Soc. *A*, 658 (1968).
214) Herberich, G. E., Jackson, R. H., Millen, D. J.: J. Chem. Soc. *A*, 336 (1966).
215) Mirri, A. M., Scappini, F., Cazzoli, J.: J. Mol. Spectroscopy *38*, 218 (1971).
216) Rigden, J. S., Butcher, S. S.: J. Chem. Phys. *40*, 2109 (1964).
217) Beagley, B.: Trans. Far. Soc. *61*, 1821 (1965).
218) Clark, A. H., Beagley, B., Cruickshank, D. W. J., Hewitt, T. G.: J. Chem. Soc. *A*, 1613 (1970).
219) Shibata, S., Bartell, L. S.: J. Chem. Phys. *42*, 1147 (1965).
220) Derissen, J. L.: J. Mol. Struct. *7*, 67 (1971).
221) Vledder, H. J., Mijlhoff, F. C., Leyte, J. C., Romers, C.: J. Mol. Struct. *7*, 421 (1971).
222) Almenningen, A., Strand, T. G.: private communication.
223) Brendhaugen, K., Fikke, M., Seip, H. M.: Acta Chem. Scand. *27*, 1101 (1973).
224) Hedberg, K.: Private communication.
225) Lox, A. P., Peneros, J. M.: J. Chem. Phys. *42*, 3106 (1965).
226) Amano, T., Saito, S., Hirota, E., Morino, Y., Johnson, D. R., Powell, F. X.: J. Mol. Spectrosc. *30*, 275 (1969).
227) Clark, A. H., Beagley, B.: J. Chem. Soc. *A*, 46 (1970).
228) Clark, A. H.: J. Mol. Struct. *7*, 485 (1971).
229) Parent, C. R., Gerry, M. C. L.: J. Mol. Spectroscopy *49*, 343 (1974).
230) Clark, A. H., Beagley, B., Cruickshank, D. W. J., Hewitt, T. G.: J. Chem. Soc. *A*, 872 (1970).
231) McDonald, W. S., Cruickshank, D. W. J.: Acta Cryst. *22*, 37 (1967).
232) Porai-Koshits, M. A., Kukina, G. A.: The Results of Science. Crystallochemistry (Russian) *9*, 6–103 (1974).
233) Hartley, F. R.: Chem. Soc. Rev. *2*, 163 (1973).
234) Appleton, T. G., Clark, H. C., Manzer, L. E.: Coordination Chem. Rev. *10*, 335 (1973).

Received July 15, 1974

The Geometric and Dynamic Structures of Fluorocarbons and Related Compounds

Dr. Akimichi Yokozeki and Prof. Simon H. Bauer

Department of Chemistry, Cornell University, Ithaca, New York, U.S.A.

Contents

I.	A. Introduction	72
	B. Fluorine for Hydrogen Substitution Effects in Di- and Tri-atomic Species	73
II.	Carbon-Fluorine Bonds	75
	A. 1. Mono-fluorocarbons: C(sp^3)–F	77
	2. Mono-fluorocarbons: C(sp^2)–F	78
	3. Mono-fluorocarbons: C(sp)–F	80
	4. Summary of the C–F Distance in the Mono-fluorocarbons	81
	B. Gem-difluorocarbons	82
	C. Tri- and Tetra-fluorocarbons: $-CF_3$ and CF_4	85
	D. Summary of Structural Data on C–F Bonds in Fluorocarbons	88
III.	Fluorine Substitution Effects	89
	A. 1. Change in Molecular Structure Adjacent to F Substitution Sites	89
	2. The Effects of Substitution of CH_3 by CF_3	89
	3. Substitution Effects in X–C(=O)R Compounds	91
	4. C–C, C=C, and C≡C Distances in Fluorocarbons	95
	5. Change of Ring Size Upon F for H Replacement	99
	B. 1. Conformations – General Remarks	101
	2. Fluoroethanes and Related Compounds	102
	3. Comments on Individual Molecules	104
	1,2-Dihaloethylenes	104
	The Azocompounds: XN=NY	105
	Propenes and Acetaldehydes	107
	1,3-Butadienes and Related Compounds	108
	Acetylacetones	109
IV.	Selected Inorganic Fluorides	111
V.	Conclusions	113
VI.	References	114

I. A. Introduction

There are several objectives of a complete molecular structure determination[1]. The first is to locate precisely in geometric space and for the lowest electronic state the deepest minimum in the potential energy surface that specifies the energy of the groups of atoms which comprise the molecular unit as a function of their internuclear distances. The second is to measure the directional curvatures at this minimum. In addition, it is interesting to explore this multi-dimensional surface in the vicinity of the deepest minimum for the location of adjacent minima, and to ascertain the heights of the barriers which separate them. The solution of this quantum mechanical problem for polyatomic molecules which incorporate a dozen or so atoms is within current capabilities of large computers if one is satisfied with a modest level of accuracy. However, the number of such computations that can be made at a level useful to practicing chemist is still severely limited, particularly by the rather extensive parameterization that must be incorporated in these computations. It is evident that for some years to come the ever growing powerful computers will not be able to overtake the precision with which interatomic distances and bond angles can be measured by diffraction and spectroscopic techniques. One approach that has received insufficient attention is the possibility that structural *differences* between similar molecules within one family could be estimated more accurately than their absolute magnitudes. It is the hope of the authors that the following review will focus attention on the wealth of experimental structural data available relative to the effects of fluorine for hydrogen substitution, for the hydrocarbons and related substances, and thus provide an incentive for theoretical studies of these differences.

Experimental techniques for determining internuclear separations in the gas phase fall into two categories:

electron diffraction and

rotational and/or vibration-rotation spectra derived from microwave absorptions, Raman and infrared spectral records.

The magnitudes of structural parameters derived from electron diffraction patterns differ from those deduced from spectroscopic data because the two types of experiments measure different averages over thermally equilibrated assemblies of molecules, which are undergoing anharmonic oscillations, with vibration-rotation interactions, etc. However, they are interconvertible on the basis of the theory of molecular dynamics. The combined use of diffraction and spectroscopic information makes possible the determination of bond lengths with a precision of 0.001Å, for selected small molecules[2]. Several reviews on structure determinations by electron diffraction techniques have appeared during the past three years.[1–5] Experimental and theoretical details are given in these papers; they also incorporate surveys of molecular structures for various classes of compounds. Concurrently, during the past five years molecular orbital studies, using both semi-empirical and *ab initio* procedures, have attained a significant level of quantitative performance.[6] However, for molecules that incorporate fluorine atoms, substantial discrepancies appear between the observed geometries and the minimum energy structures.

In this review we shall first survey the variety of C–F bonds that have been investigated. We will then consider the effect of fluorine for hydrogen substitution on the rest of the molecule. The third general topic concerns intramolecular motions and related internal rotations about single bonds.

Most of the structural parameters in this review were derived from electron diffraction data. Clearly, precise comparisons of structures should be made on the basis of the same "measure" of internuclear distances. Because of the absence of stable fluorine isotopes microwave data on the fluorocarbons are generally incomplete, particularly for structures that are characterized by a significant number of parameters. To avoid ambiguity and confusion the following symbols for the different measures of distance were used (see Ref.[7]) for a more complete discussion):

- r_e Distance between nuclear positions at the minimum in the potential energy function, $V(r_{ij})$.
- r_g Thermal-average value of internuclear distance (electron diffraction)
- r_a Value at the center of gravity of the distribution function $P(r)/r$, which is convenient to evaluate from electron diffraction data; $P(r)$ is the distance distribution function.
- r_0 "Effective" distance, derived directly from ground-state rotational constants.
- r_s "Effective" distance derived from rotational constants *via* Kraitchman's equations for a sequence of isotopic substitutions.

The r_s distance is believed to be close to the r_e; roughly speaking, the difference between them is at most 0.005 Å. Differences between r_s and r_g are generally of the order of 0.01Å ($r_g > r_s$). The r_a and r_g measures of distance are related by the approximate equation $r_g \simeq r_a + l^2/r_a$ (l is the root mean square amplitude of vibration). For example,
r_g(C–H) $\cong r_a$ (C–H) + 0.005_5 Å;
r_g (C–C) $\cong r_a$ (C–C) + 0.001_8 Å, and
r_g (C–F) $\cong r_a$ (C–F) + 0.001_5 Å.
Parallel to the different measures of internuclear distance there are corresponding measures of bond angles. The differences between these are usually small compared with the current experimental uncertainties.[7] Hence, in this article they are used without discrimination.

I. B. Fluorine for Hydrogen Substitution Effects in Di- and Tri-atomic Species

Casual inspection of the limited structural information[8,9,10] available for di and triatomic hydrides and fluorides shows that for these small molecules the differences are chracteristic and follow a pattern which reappears in comparisons of more complex hydrides and fluorides. Let

$$\delta R \equiv r_e\,(\text{M–F}) - r_e\,(\text{M–H}).$$

Then, for M = Li, Be, B, C, N, O and F,
δR is –0.03, 0.02, 0.03, 0.15, 0.27, (0.38) and 0.50 Å,
respectively. The magnitudes of the differences are relatively constant (± 0.03 Å) from Li to B, and increase linearly with group number across the periodic table. The second row elements show a similar trend (Fig. 1 and Table 1a). These sequences may be accounted for approximately, in terms of the Schomaker-Stevenson equation:

$$r(\text{M–X}) = r_\text{M} + r_\text{X} - \beta\,|\chi_\text{M} - \chi_\text{X}|.$$

δR is given by $\delta R = 0.18$ Å, for $\chi_M \leqq 2.1$, and $\delta R = 0.18\,\chi_M - 0.20$ (Å), for $\chi_M > 2.1$,

where χ_M is the electronegativity of the M atom. For the hydrides of the first row elements there is an inverse correlation between r_e (M–H) and D_0 (M–H), the bond dissociation energy, and there is a corresponding relation for the fluorides.

There are about a dozen tri- and tetra-atomic species for which trends are discernible upon F/H substitution. First, r(Q–F) is less in QF_2 species than in QHF; second, r(Q–H) is less in QH_2 species than in QHF. It will become apparent from the following compilations

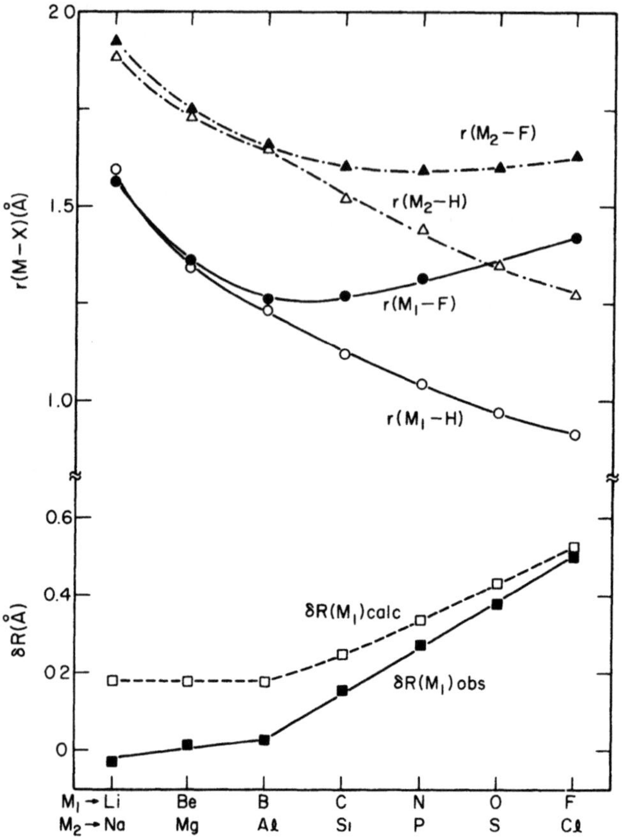

Fig. 1. Internuclear separation for the fluorides and hydrides of the elements in the first and second rows of the periodic table (upper curves). The lower set compares the estimated differences [δR (M_1)] with those observed for the first row elements; see Text

Table 1a. Internuclear distances in diatomic hydrides and fluorides[1])

	Li	Be	B	C	N	O	F
H	1.5953	1.3431	1.236	1.1202	1.045*	0.9706	0.9170
F	1.5639	1.361	1.2626*	1.2718	1.3173	[1.35][2])	1.418*
	Na	Mg	Al	Si	P	S	Cl
H	1.8873	1.7306	1.6482	1.521	1.4388*	1.3503*	1.2745
F	1.9259	1.7500	1.6544	1.6008	1.5896	1.599*	1.6281

[1]) r_e (Å) distances except for those designated by * (r_0): Data taken from Refs. [1, 8, 10]. (Lowest electronic states.)

[2]) Estimated from Fig. 1.

Table 1b. Distances (Å) and bond angles in triatomic hydrides and fluorides[1])

	C–H	C–F	Bond angle	N–H	N–F	Bond angle	O–H	O–F	Bond angle
H_2Q	1.11	–	104°	1.026	–	103.3°	0.958	–	104.9°
HQF	1.12	1.314	101.8°	1.06	1.37	105°	0.964	1.442	97.2°
QF_2	–	1.30_0	104.9°	–	1.363	102.5°	–	1.409	103.3°

[1]) Structures in the lowest electronic states.

$CH_2(^1A_1)$, $CHF(^1A')$, $NH_2(^2A_1)$, $NHF(^2A'')$: Ref. [9]).
CF_2: Powell, F. X., and Lide, D. R., Jr.: J. Chem. Phys., *45*, 1067 (1966); Mathews, C. W.: ibid., *45*, 1068 (1966).
NF_2: Bohn, R. K., and Bauer, S. H.: Inorg. Chem., *6*, 304 (1967).
H_2O: Laurie, V. W., and Herschbach, D. R.: J. Chem. Phys., *37*, 1687 (1962); Shibata, S., and Bartell, L. S.: ibid., *42*, 1147 (1965).
HFO: Kim, H., Pearson, E. F., and Appelman, E. H.: J. Chem. Phys., *56*, 1 (1972).
F_2O: Pierce, L., Jackson, R. H., and DiCianni, N.: J. Chem. Phys., *35*, 2240 (1961).

of structural data that these trends are general. For example, in NH_3[8]), NHF_2[11]), NF_3[12]), the N–H and N–F separations are: (1.012, ____), (1.026, 1.400), and (____, 1.365) Å; in the sequence H_2CO[13]), HFCO[14]) and F_2CO[15]), the C–H, C–F and C=O bond lengths are, respectively: (1.101, ____, 1.203), (1.095, 1.338, 1.181), and (____, 1.317, 1.170), Å. Similarly, in H_2CH_2, $HFCH_2$ and F_2CH_2, the C–H and C–F distances[8]) are, respectively: (1.0938, ____), (1.106, 1.385), and (1.092, 1.358), Å.

II. Carbon-Fluorine Bonds

In 1937 Brockway reported that C–F bond distances in the fluoromethanes and in the fluorochloromethanes were significantly shorter in those compounds where several fluorine atoms are attached to the same carbon atom than in the monofluorides.[16]) Since then many proposals have been presented to account for this progressive bond shortening with increasing fluorination. In terms of valence bond theory[17]) the effect was correlated with changes in hybridization and electronegativity.[18]) One of the earliest explanations was formulated as double bond-no bond resonance.[17,19])

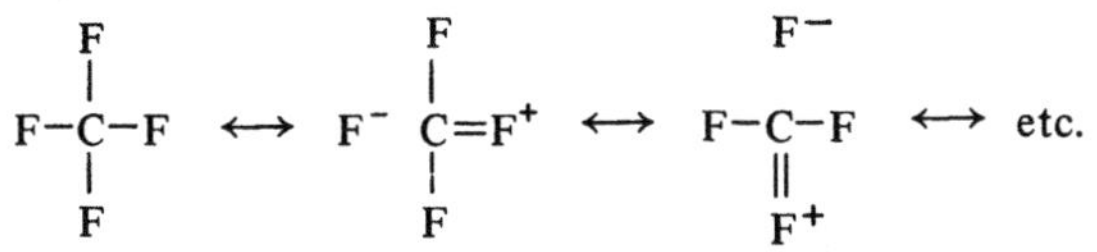

Subsequently, many examples of this "fluorine effect" were discovered not only for more complex fluorocarbons, but also for the other similarly coordinated elements: Si, B, P, etc.[17]) It was also demonstrated that whereas other halogen derivatives show a similar progressive bond shortening with increasing halogenation, the magnitude of the effect is much smaller, as illustrated in Fig. 2. For Br and I, bond shortening is not discernible within the experimental uncertainties.[8]) The contrast between the fluorine and the other

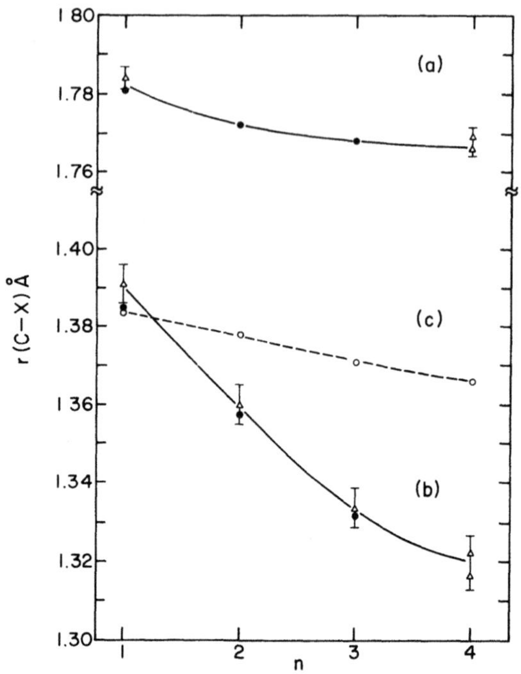

Fig. 2. C–X distances in $CH_{4-n}X_n$. (*a*) X = Cl; (*b*) X = F.

: Designates electron diffraction r_a's with experimental uncertainties.

: Designates r_0 or r_s from microwave data.

Most data are from Ref. [8]. For CCl_4: Morino, Y., Nakamura, Y., and Iijima, T.: J. Chem. Phys., *32*, 643 (1960); Haase, J., and Zeil, W.: Z. Phys. Chem., *45*, 202 (1965). For CH_3Cl: Bartell, L. S., and Brockway, L. O.: J. Chem. Phys., *23*, 1860 (1955).

(*c*) – –O– –; calculated values, SCF MO, by Pople, *et al.*[21]

Table 2. Correlation of molecular parameter for the fluoromethanes

	r (C–F)[1]	*E* (C–F)[2]	*ν* (C–F)[3]	*K* (C–F)[4]	*J* (CH)[5]
Unit	Å	kcal/mole	cm^{-1}	md/Å	Hz
CH_3F	1.391 (1.385)	107	1049	3.71	149.0
CH_2F_2	1.360 (1.358)	109.6	1084	–	184.5
CHF_3	1.334 (1.332)	114.6	1140	–	239.1
CF_4	1.323 or 1.317	116	1188	4.18	–

1) Ref. [8] (r_a and r_0 in parentheses).

2) Bond energy: Patrick, C. R.: Advances in fluorine chemistry, Vol. 2, Chapter 1. Washington: Butterworths 1961.

3) Degeneracy-weighted averages of symmetric and antisymmetric C–F stretching frequencies taken from Shimanouchi, T.: Tables of molecular vibrational frequencies, National Bureau of Standards, and Pace, E. L.: J. Chem. Phys., *18*, 881 (1950).

4) Stretching force constants based on the Urey-Bradley force field: Shimanouchi, T.: Physical chemistry, an advanced treatise, Vol. 4, Chapter 6, 233 (1970).

5) ^{13}C–H coupling constants: Levy, G. C., and Nelson, G. L.: Carbon-13 nuclear magnetic resonance for organic chemistry. New York: John Wiley and Sons 1972.

halogen derivatives has indeed stimulated widespread studies, both theoretical and experimental, of "fluorine substitution effects". Current developments in molecular orbital theory provides a base for a quantitative analysis, but even the most sophisticated attempts[20,21] have not yet adequately accounted for the observed increments, although the progressive shortening was found, as shown in Fig. 2.

As is well known, a full description of chemical bonding covers many molecular parameters such as bond energies, stretching frequencies, force constants and J_{CH} ($^{13}C-H$ coupling constants); these complement the magnitudes assigned to mean internuclear separations. Values in Table 2 illustrate progressive changes with increasing fluorination. However, to date the correlation between them is only empirical [distance-force constant equations (Badger[22]), distance-bond energy relations (Dewar, *et al.*[23]) and C–H distance-J_{CH} (Muller, *et al.*[24])].

In the subsequent sections, the various types of C–F bond lengths will be summarized according to the number of fluorine atoms attached to a given carbon atom.

A.1. Mono-fluorocarbons: C(sp^3)-F

Mean internuclear carbon-fluorine distances in mono-fluorocarbons may be classified according to whether the carbon atom is primary, secondary, or tertiary (Table 3).[25–36] Within each type, the reported C–F distances range over 0.01 Å, partly because different measures of "mean distance" were used by the investigators and partly due to intrinsic factors. With this in mind one may choose typical values for each type: 1.39 Å (primary), 1.41 Å (secondary) and 1.43 Å (tertiary).

In spite of the absence of many examples, the ascending sequence for primary, secondary and tertiary carbon-halogen bond lengths is apparent; thus, for C–Cl bonds,
CH_3CH_2Cl (r_s = 1.788 ± 0.002 Å),[37]
$(CH_3)_2CHCl$ (r_s = 1.798 + 0.008/–0.004 Å[38]) and
$(CH_3)_3CCl$ (r_s = 1.803 ± 0.002 Å)[39];

Table 3. C–F distances (Å) in monofluorocarbons, for C (sp^3)

Type	Typical value	Example
Primary –CH_2F	1.39	FCH_3 (r_a = 1.391 ± 0.005[25]; r_s = 1.385[26]) FCH_2CH_3 (r_s = 1.398 ± 0.005)[27] FCH_2COOH (r_a = 1.387 ± 0.010)[28] $FCH_2CH{=}CH_2$ [r_s = 1.382 ± 0.010 (*cis*), 1.371 ± 0.015 (*gauche*)][29] FCH_2CH_2OH (r_a = 1.400 ± 0.005)[30] FCH_2CH_2F (r_a = 1.389 ± 0.007[31]; r_a = 1.394 ± 0.001[32])
Secondary >CHF	1.41	Monofluorocyclohexane (r_a = 1.41)[33] CH_3CHFCH_3 (r_g = 1.408 ± 0.002)[34]
Tertiary ≥CF	1.43	$FC(CH_3)_3$ (r_a = 1.425 ± 0.008;[35] r_0 = 1.43 ± 0.02[36])

for C-Br bonds, CH_3CH_2Br (r_s = 1.950 ± 0.004 Å)[40] and
$(CH_3)_2CHBr$ (r_s = 1.957 + 0.011/−0.004 Å[41]);
for C–H bonds, CH_3CH_3 (r_s = 1.095 ± 0.002 Å[42], r_g = 1.111 ± 0.002 Å)[43] and
$(CH_3)_3CH$ (r_s = 1.108 ± 0.001 Å[44], r_g = 1.122 ± 0.006 Å)[45].

However, this classification does not apply to "strained ring" compounds because geometrical requirements imposed by ring closure probably affect the state of carbon hybridization. For example, the tertiary C–Cl distance in 4-chloronortricyclene and 1,4-dichloronorbornane are 1.763 ± 0.003 Å (r_g)[46] and 1.773 ± 0.008 Å (r_g)[47], respectively. They are definitely shorter than the corresponding value in t-butylchloride (r_g = 1.828 ± 0.005 Å)[48]. The C–Cl distance (r_s) of 1.740 Å in monochlorocyclopropane[49] is about 0.04 Å shorter than in monochloromethane[26]. One may anticipate that corresponding C–F distances in cyclic compounds would also be shorter than in the chain compounds; no data for this category are available. Additional discussions of the "ring effect" are presented in section II.B.

A.2. Mono-fluorocarbons: C(sp^2)–F

It has long been recognized that because of conjugation a C–F bond adjacent to a double bond is shorter than one attached to a saturated carbon atom. Parenthetically, the notations used for hybridization: sp^3, sp^2 and sp are merely convenient designations, since a range of mixed states must be considered. Several examples of compounds that incorporate C(sp^2)–F bonds are listed in Table 4.[14,50–56] There is a spread of 0.01 Å about the

Table 4. C–F distances in monofluorocarbons, for C (sp^2)

Compounds	r (C–F) Å	
F–CH=CH_2	1.348 ± 0.004[50]	(r_g)
	1.348[51]	(r_s)
F–CH=$CHCH_3$	1.342 ± 0.007[52,53]	(r_o)
F–CHO	1.351 ± 0.013[54]	(r_a)
	1.338 ± 0.005[14]	(r_s)
F–$COCH_3$	1.348[55]	(r_s)
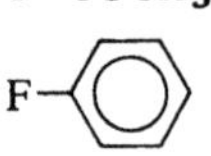	1.354[56]	(r_s)
Typical value	1.35	

typical value of 1.35 Å, indicative of a shrinkage of ≈ 0.04 Å in =CHF compared with $-CH_2F$. In addition to the influence of the adjacent multiple bond, a "secondary environment effect" is introduced when the multiple bond ends in a heteroatom such as oxygen or nitrogen. The C–F bond in F–CHO[14] is about 0.01 Å shorter (r_s) than in F–CH=CH_2[51]. In F–$COCH_3$[55], however, such a trend is not apparent from the data now available (Table 4). When a fluorine atom is substituted onto a β-carbon, in ethylenefluoride, a secondary environment effect appears as a decrease in the C(α)–F distance; in FCH=CHF [(C–F) = 1.344 ± .002 Å (*trans*) and 1.335 ± .002 Å (*cis*); r_g].[50] These are 0.004 and 0.013 Å shorter, for the *trans* and *cis* forms, respectively, than in $FCH{=}CH_2$ (1.348 ± .004 Å; r_g).[50] In this connection the most striking effect on C–F produced by fluorine for hydrogen substitution onto an adjacent or next to adjacent carbon atom was reported for the fluorobenzenes (Table 5). Although data are not available for all combinations of substitution, interactions between fluorine atoms on different carbon atoms are clearly indicated. Compared with monofluorobenzene,[56] the other fluorobenzenes[57–63] have shorter C–F distances.

Table 5. C–F bond distances in the fluorobenzenes (Å)

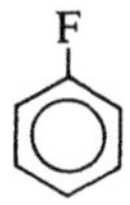	r_o 1.354	Nygaard, *et al.* (1968)[56]
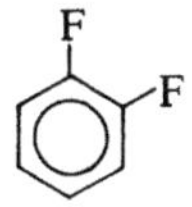	r_o 1.294 – 1.313 r_o 1.311 – 1.318	Hatta, *et al.* (1968)[57] Nygaard, *et al.* (1967)[58]
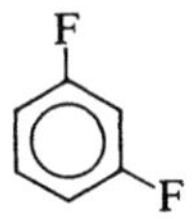	r_o 1.301 – 1.308 r_g 1.324 ± .002	Nygaard, *et al.* (1967)[58] Schaick, *et al.* (1973)[59]
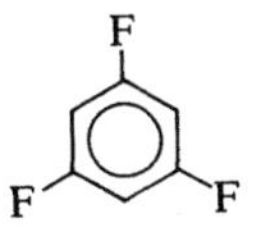	r_g 1.305 ± .010 r_o 1.304	Bauer, *et al.* (1968)[60] Schlupf and Weber (1973)[61]
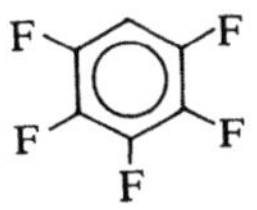	r_o 1.328	Doraiswamy and Sharma[62]
	r_g 1.324 ± .006 r_a 1.327 ± .007 r_o 1.321	Bauer, *et al.* (1968)[60] Almenningen, *et al.* (1964)[63] Schlupf and Weber (1973)[61]

The *ortho* and *meta* effects are prominent; the respective C–F distances are 1.306 Å (with r_0 assigned to the mean of 1.294–1.318 Å) in *o*-difluorobenzene[57,58] and 1.304 Å (r_0 assigned to the mean of 1.301–1.308 Å) in *m*-difluorobenzene[58]. It is tempting to classify the C–F bond distances in the fluorobenzenes according to three geometric types:

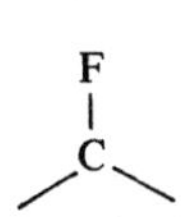
(I)
(r_0) 1.354 Å

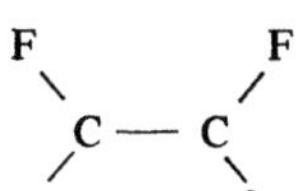
(U)
1.306 Å

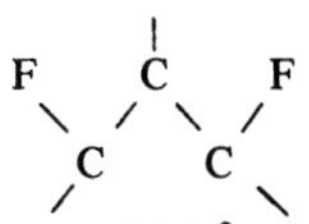
(W)
1.304 Å

This is parallel to the classification proposed for the fluoroethylenes by Epiotis[64] based on molecular orbital considerations. He postulated three contributing factors: charge transfer through space, charge transfer through bonds, and direct orbital interactions. He found strong attractive interactions for U-type

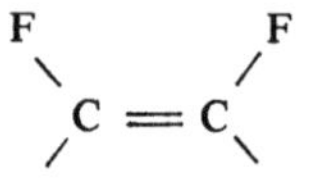

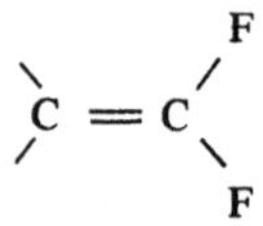

geometries, and weak attractive interactions for the Y-type configurations.

Let us postulate that the three types (I, U and W) are combinatorial elements which determine the C–F bond distances in the fluorobenzenes, and compute the mean of all interactions weighted by the number of contributing terms. Then in penta- and hexa-fluorobenzenes the C–F distances are given by
$[5r(\mathrm{I}) + 4r(\mathrm{U}) + 4r(\mathrm{W})]/13 = 1.324$ Å and
$[6r(\mathrm{I}) + 6r(\mathrm{U}) + 6r(\mathrm{W})]/18 = 1.321$ Å,
respectively. In terms of r_0 these prove to be good estimates for the corresponding observed values (1.328[62] and 1.321 Å[61]). Even if we accept the usefulness of such an empirical approach, more accurate C–F distances for these fluorobenzenes, including other fluoro substituents, should be obtained. For instance, the C–F distances in *m*-difluorobenzene, 1.324 ± .002 Å, reported in a recent electron diffraction study by Schaick, *et al.*[59] are about 0.02 Å longer than the corresponding value derived from spectroscopic data (1.301–1.308 Å)[58]; the difference is much larger than could arise from differing measures of mean distances.

Because of strong correlations between nearly equal bond distances the least squares reduction for diffraction data makes it difficult to separate the C–F distance from the C–C distances in benzene rings; indeed, some combinations are inherently unresolvable. On the other hand, spectroscopic techniques are also not free from ambiguity because of the lack of more than a single stable F isotope. One must assume some elements of symmetry and/or C–C distances in the rings to derive C–F bond lengths. A combination of diffraction and spectroscopic data may provide more accurate values for the C–F bonds of fluorobenzenes. Because the electron diffraction value for *m*-difluorobenzene does not fit the IUW pattern, consider the *weighted average* of C–C and C–F bond distances derived from the diffraction data;[59] it is 1.373 Å. If we assume no ring distortion and a C–C distance of 1.396 or 1.400 Å, as was done in the spectroscopic study[58], the reverse calculation for the corresponding C–F distance gives 1.304 or 1.292 Å, which is compatible with the spectroscopic estimate (1.301–1.308 Å).[58]

A.3. Mono-fluorocarbons: C(*sp*)–F

For the C–F bond distances in ≡C–F types, only (r_s) microwave data are available (Table 6)[65]. The difference between these C–F distances may be ascribed to a "secondary environmental effect", one that should be predicted by an adequate MO calculation. A similar but much smaller trend was reported for other X–C≡CH and X–C≡N pairs; when X is CH_3, CH_3–C≡CH (r_g = 1.470 ± 0.004 Å)[2] and CH_3–C≡N (r_g = 1.466 ± 0.003 Å),[2, 66] and when X is Cl, Cl–C≡CH (r_s = 1.637 Å)[65] and Cl–C≡N (r_s = 1.631 Å).[65] For contrast, note that the C–H bond distance in H–C≡CH is about

Table 6. C (sp)–F Distances

Compounds	C–F (Å)
F–C≡CH[65]	1.279 (r_s)
F–C≡N[65]	1.262 (r_s)

0.011 Å *shorter* than in HCN; r_0 = 1.0566 ± 0.0003 Å[67] in the former, but r_0 = 1.0676 Å[68] in the latter. Similarly, when X is H_2C=CH– as in vinyl acetylene, the C–C distance appears to be slightly shorter than in vinyl cyanide: H_2C=CH–C≡CH (r_g = 1.434 ± 0.003 Å[69]) compared with (r_g = 1.438 ± 0.003 Å[70]) in H_2C=CH–C≡N.

A.4. Summary of the C–F Distance in the Mono-fluorocarbons

The C–F bond distances in the monofluorinated species follow a general pattern in that their magnitude decreases with decreasing n, for sp^n carbon atom hybridization. The effects of conjugation due to adjacent multiple bonds are shown in Table 7. Similar

Table 7. C–F bond distances adjacent to variously bonded carbon atoms

Type	C–F (Å)	Δ[1] (%)
$-CH_2F$	1.39	–
=CHF	1.35	2.9
≡CF	1.27	8.6

[1] Referred to the C–F bond length in $-CH_2F$; see see Text for the definition of Δ.

Table 8. Comparison of Δ[1] (%)

X	=CH–X	≡C–X
F	2.9	8.6
Cl	3.1	8.2
Br	2.5	7.7
I	2.0	6.9
H	0.7	3.0
$-CH_3$	1.8	4.3
–CHO	2.0	4.0
$-CH=CH_2$	2.7	4.7
–C≡CH	2.5	5.7
–C≡N	1.9	5.2

[1] Relative C–X bond distances in CH_2=CHX, and in HC≡CX (or N≡CX), referred to the C–X distance in CH_3–X: see Text for the definition of Δ. The corresponding distances were taken from Refs.[2, 8, 65, 110, 162–164].

effects on neighboring C–X distances, where X is a halogen or another functional group, are illustrated in Table 8. It appears that C–X bond distances are affected by (*sp*) hybridization to the extent of 2–3% for all =CHX cases except when X is H. Define

$$\Delta \equiv [r(C_{sp^3}\text{-}X) - r(C_{sp^n}\text{-}X)]/r(C_{sp^3}\text{-}X),\ n = 1,2.$$

Then for ≡C–X the Δ value amounts to 7~9% when X is a halogen, and 4~6% when X is $-CH_3$, –CHO, $-CH{=}CH_2$, –C≡CH and –C≡CN. C–H distances are least affected by the change in carbon hybridization. Thus, when viewed in context of other conjugating groups, the C–F bond distances are not "abnormal", although the magnitudes of their Δ's are larger than for the others.

Apart from the above "primary environment effect", a "secondary environment effect" appears for C–X in multiply bonded cases such as Y=CHX and Y≡CX (where Y is O, N, CHF, etc.). Among them the largest effect was observed in the fluorobenzene and in FC≡N.

II. B. Gem-difluorocarbons

There are two structural types in this category: $C(sp^3)F_2$ and $C(sp^2)F_2$ (Table 9). To date a few examples have been investigated; the fluorocycloalkanes which belong to the first type are discussed in the latter part of this section.

Table 9. Typical distance in gem-difluorocarbons (Å)

Type	C–F	Examples
$>CF_2$	1.36	CH_2F_2 (r_o = 1.358 ± 0.001[1]), r_a = 1.360 ± 0.005[25])
$=CF_2$	1.32	$CH_2{=}CF_2$ (r_g = 1.324 ± 0.003[50]), r_a = 1.321[2]), r_s = 1.323[72])
		$F_2C{=}O$ (r_s = 1.312 ± 0.010[71]), r_z = 1.316_6 ± 0.001[15])

1) Lide, D. R., Jr.: J. Am. Chem. Soc., *74*, 3548 (1952).
2) Karle, I. L., and Karle, J.: J. Chem. Phys., *18*, 936 (1950).

Table 10. Comparison of bond angles in analogous compounds

X	H_2CX_2 (∠XCX)	$H_2C{=}CX_2$ (∠XCX)	$O{=}CX_2$ (∠XCX)
F	108.3° ± 0.1°[1])	109.3° ± 0.6°[50])	108.0° ± 0.5°[71])
H	109.5°	116.2° ± 0.8°[161])	116.5°[13])
CH_3	112.4° ± 1.2°[157])	115.8° ± 0.6°	116.7° ± 0.3°[108])
Cl	~ 113°[8])	~ 114°[8])	111.3° ± 0.1°[2])

1) Lide, D. R., Jr.: J. Am. Chem. Soc., *74*, 3548 (1952).
2) Robinson, G. W.: J. Chem. Phys., *21*, 1741 (1953).

Parallel to the monofluorocarbons, the gem-difluoro compounds also have shorter C–F bonds for carbons hybridized with smaller *n*'s: from 1.36 Å [$C(sp^3)F_2$] to 1.32 Å [$C(sp^2)F_2$]. The effect of adjacent heteroatoms can be seen in $F_2C{=}O$ [r_s (C–F) = 1.312 ± 0.010 Å[71]] and $F_2C{=}CF_2$ [r_g (C–F) = 1.319 ± 0.002 Å[50]] compared with $CH_2{=}CF_2$ (r_g = 1.324 ± 0.003 Å[50]), r_s = 1.323 Å[72]); CCl_2F_2 (r_a = 1.338 ± 0.013 Å[73]) compared with CH_2F_2 (r_a = 1.360 ± 0.005 Å[25]). Note that the differences are barely significant when compared with the experimental error. One should also note that the

Table 11. Geminal C–F bonds in typical cycloalkanes, C_nX_{2n} (n = 3, . . . , 6), compared with hydrogen analogs[1])

n	r_g (C–F), Å	∠ FCF	r_g (C–H), Å	∠ HCH
3	1.314 ± 0.001[76]	112.2° ± 1.0°	1.094 ± 0.003[79]	115.1° ± 1.0°
4	1.333 ± 0.002[77]	109.9° ± 0.3°	1.098 ± 0.01[80]	–
5	–	–	1.113 ± 0.002[81]	–
6	1.339 ± 0.002[78]	109.1° ± 0.3°	1.121 ± 0.004[82]	107.5° ± 1.5°

[1]) The r_g distances for n = 3, 4, and 6 were estimated from the reported r_a structures.

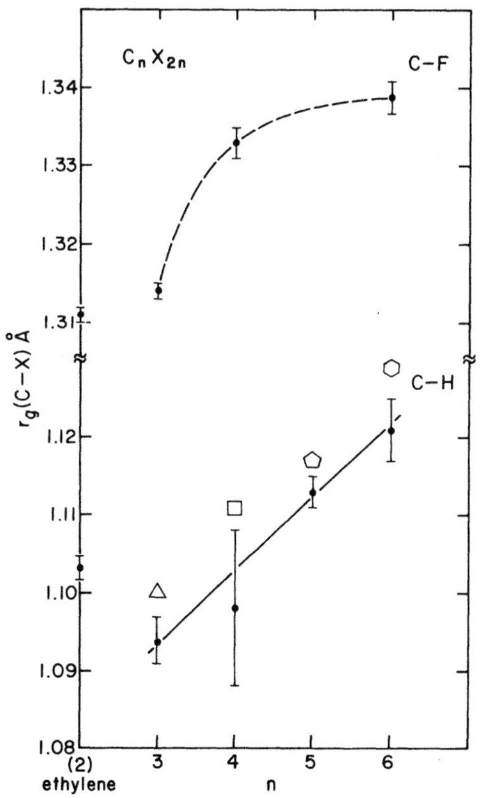

Fig. 3. C–X distances for C_nX_{2n} cyclic compounds, as a function of ring size.

bond angles between the C–F bonds are considerably smaller than those expected for sp^3 or sp^2 hybridization (Table 10). In $=CF_2$ the FCF angles are close to tetrahedral, being 7~8° smaller than the corresponding HCH or CCC angles. The "closing up" of the FCF angles have been discussed qualitatively on the basis of HAO (hybrid atomic orbitals)[18] and VSEPR (valence-shell electron-pair repulsion) theories.[74,75]

The C–F bond distances in the perfluorocycloalkanes (C_nF_{2n})[76–78] are summarized in Table 11, where they are compared with increments in corresponding C–H distances;[79-82] the C–F distances increase with ring size from 1.314 ± 0.001 Å[76] for $n = 3$ to 1.339 ± 0.002 Å[78] for $n = 6$. A similar trend appears in the C–H distances, as illustrated in Fig. 3. Surprisingly, for $n = 3 \rightarrow 6$ the latter follow a linear relation,

$$r_g\,(\text{C–H}) = 1.067 + 0.009\,n,\ \text{Å} \qquad (n = 3,4,5,6)\,.$$

For C–F only three points are available and through these a straight line cannot be drawn to connect them within the stated error bars. Indeed, they follow the expected "roll-off". On this basis one may predict that in perfluorocyclopentane ($n = 5$) the C–F distance is 1.335 ± 0.010 Å.

Attention is called to another interesting feature of this series. The increment in the C–H distance associated with ring size (0.027 Å for $n = 3$ to $n = 6$) is comparable in magnitude to the difference between the C–H bond lengths attached to sp^3 vs sp carbons: (0.033 Å; from 1.111 ± 0.002 Å[43] in CH_3–CH_3 to 1.078 ± 0.002 Å[2] in CH≡CH). This is suggestive of a parallel trend in sp^n hybridization of carbon from cyclohexane to cyclopropane. Walsh[83] postulated for cyclopropane sp^2 hybridization for the bonds external to the cyclopropane ring and a mixture of p and sp^2 orbitals for the internal bonds. The net effect appears to have reduced the C–H bond length in cyclo-C_3H_8 below that in C_2H_4. Further, the NMR ^{13}C–H coupling constant (J_{CH}), which is a rough measure of the s character associated with the carbon atom[84–86], parallels the sequence illustrated in Fig. 2: cyclopropane (161 Hz),[87] cyclobutane (134 Hz)[87] and cyclohexane (123 Hz)[87]. J_{CH} in cyclopropane is slightly larger than those for ethylene (156.2 Hz)[88] and benzene (158.5 Hz)[89], whereas J_{CH} in cyclohexane is in the vicinity of the magnitudes for methane (125 Hz)[24] and ethane (124.9 Hz)[88]. Similarly, the carbon fluorine coupling constants in cyclic geminal difluorides appear to be sensitive to the hybridization of the carbon; $|J_{CF}|$ becomes larger as the s character increases.[90] In gem-difluoro cyclohexane $|J_{CF}|$ is 240 Hz[90]; it is 280 Hz[90] in gem-difluorocyclobutane and 330 Hz in the perfluoropropane[91]. However, since the relation between the coupling constants and bond distances is empirical, care should be taken in extending it beyond the range of observation.

The XCX bond angle in the C_nX_{2n} appear to be close to tetrahedral for both the hydrogen and fluorine cases except for $n = 3$; then ∠FCF and ∠HCH are about 112° and 115°, respectively (Table 11). The geminal C–F (or C–H) bond lengths in typical cyclic compounds correlate with the ring size; *i.e.* with the internal bond angle

The relation between them is illustrated in Fig. 3.[76–78;92,93] The largest deviation is shown by perfluorocyclopentene.[77] One may anticipate that the geminal C–F distances in the more complicated cyclic compounds, such as bridged or fused ring molecules, would also correlate with the trend indicated in Fig. 4. In addition to the effect of ring size on the external bond lengths (C–F or C–H), the replacement of H by F affects the ring dimensions; this is discussed in a later section.

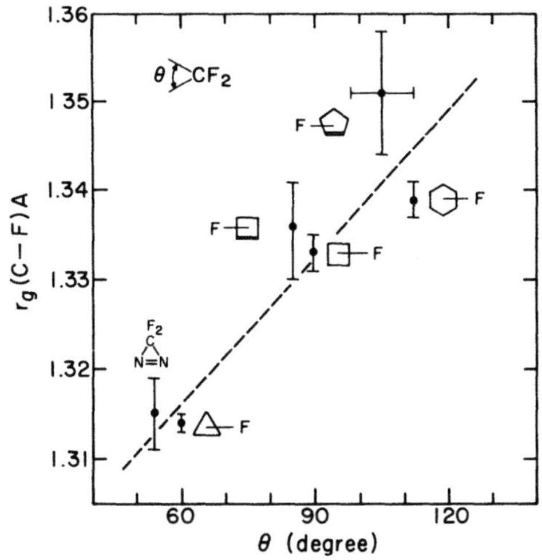

Fig. 4. The geminal C–F distances in cyclic compounds correlate with the internal bond angle. Data for: F_2CN_2 (Ref. [92]); C_3F_6 (Ref. [76]); C_4F_8 and C_4F_6 (Ref. [77]); C_6F_{12} (Ref. [78]); and C_5F_8 (Ref. [93])

II. C. Tri- and Tetra-fluorocarbons: $-CF_3$ and CF_4

Among the fluorocarbons, molecules which belong to this category have been most extensively studied.[25,35,94–121] The structure of $-CF_3$ units appears to be unperturbed by its environment: r(C–F) ≈ 1.33 Å and ∠FCF ≈ 108.5°. This generalization holds, as

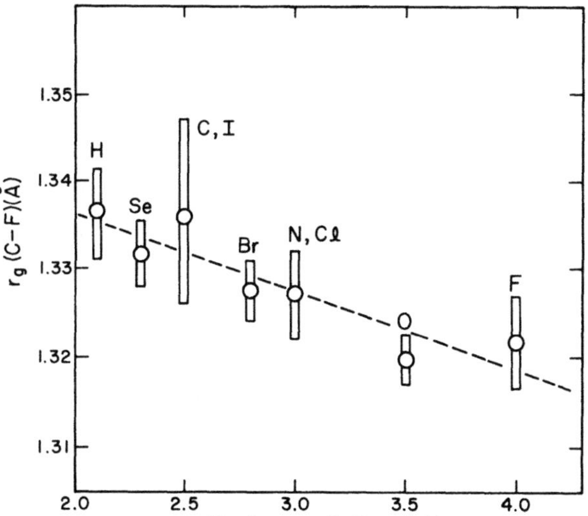

Fig. 5. The C–F distances in X–CF_3 vs the electronegativity of X. Here X is the atom attached to the CF_3 group. The vertical bars in the figure represent the scatter of experimental values as presented in Table 12, whereas the circles indicate the average values for each category of compounds

Table 12. Structures of $-CF_3$ units

Compounds	C–F, Å[1]	∠FCF[4]	Refs.
*$H-CF_3$	1.336 ± 0.005	108.5°	25)
	(r_o) 1.332 ± 0.001	108.8°	94)
CH_3-CF_3	1.346 ± 0.006	107.3° ± 1.0°	95)
	(r_o) 1.335 ± 0.005	107.9°	96)
$CH_2=CH-CF_3$	1.347 ± 0.003	106.8° ± 0.2°	97)
$CH\equiv C-CF_3$	1.337 ± 0.01	107.5° ± 1.0°	98)
$N\equiv C-CF_3$	1.335 ± 0.005	108.5° ± 0.5°	99, 100)
$CH_3N=N-CF_3$	1.332 ± 0.003[2]	109.7° ± 0.6°	101)
$CH_3C\equiv C-CF_3$	(r_o) 1.340	107.5°	102)
$CF_3C\equiv C-CF_3$	1.331 ± 0.003[2]	108.1° ± 0.3°	103, 104)
$(CF_3)_3CH$	1.336 ± 0.002	108.0° ± 0.2°	35)
$(CF_3)_3COH$	1.335 ± 0.004	108.3° ± 0.4°	105)
CHF_2-CF_3	(r_s) 1.336	108.1°	106)
CF_3-CHO	1.334 ± 0.007	108.7° ± 1.0°	95)
CH_3COCF_3	1.339 ± 0.003	108.2° ± 0.3°	107)
CF_3COCF_3	1.337 ± 0.002	108.8° ± 0.3°	108)
$(CF_3)_2Se$	1.335 ± 0.004[3]	108.8° ± 0.4°	109)
$Cl-CF_3$	1.330 ± 0.002	108.6° ± 0.4°	110)
$I-CF_3$	1.344 ± 0.004	107.6° ± 0.4°	111)
	1.332 ± 0.004	108.3° ± 1.0°	99)
$Br-CF_3$	1.327 ± 0.002	109.2° ± 0.5°	99)
$(CF_3)_2Se_2$	1.328 ± 0.005	109.9° ± 0.8°	112)
*CF_3-CF_3	1.326 ± 0.002	108.5° ± 0.1°	113)
CF_3-NO	1.326 ± 0.003	110.0° ± 0.4°	114, 115)
CF_3-NO_2	1.325 ± 0.005	110.0°	116)
$CF_3N=N-CF_3$	1.326 ± 0.002[2]	109.6° ± 0.3°	101)
$(CF_3)_2N-N(CF_3)_2$	1.325 ± 0.005	108.2° ± 0.5°	117)
$(CF_3)_2C=CH_2$	1.329 ± 0.002	108.4° ± 0.2°	108)
$(CF_3)_2C=NH$	1.326 ± 0.003	108.9° ± 0.4°	108)
$(CF_3)_2NO$	1.322 ± 0.004	109.8° ± 1.0	118)
CF_3-OF	1.319 ± 0.003	109.4° ± 0.8°	119)
*CF_3-F	1.319 ± 0.005	109.5°	120)
	1.325 ± 0.005	109.5°	25)
$(CF_3)_2O$	1.320 ± 0.002	108.8° ± 0.3°	121)
$(CF_3)_2O_2$	1.317 ± 0.002	108.8° ± 0.2°	121)
$(CF_3)_2O_3$	1.321 ± 0.002	108.6° ± 0.2°	121)
	1.336	108.3°	(Upper section)
Average[5]	1.326	109.2°	(Middle section)
	1.320	109.2°	(Lower section)

* Typical for that group of compounds.

1) Distances quoted are r_g unless indicated otherwise; r_a distances in the literature were converted to r_g.
2) Three times least squares standard deviations as reported in the literature.
3) Mean value of two models given in the literature.
4) Some of the angles were calculated from the XCF angles in the $X-CF_3$ molecules.
5) Average values (r_g) for the three sections; the average r_g (C–F) for all the molecules listed is 1.331 Å.

shown in Table 12 within uncertainties of ≈ 0.015 Å and 1.5°. However, closer inspection of the Table permits a more refined classification based on three typical C–F distances: 1.336, 1.326, and 1.320 Å (r_g), as present in the representative molecules HCF_3, CF_3–CF_3, and FCF_3, respectively. Of course, the dividing lines are fuzzy. The range in C–F distances may be ascribed to environmental factors, such as the electronegativity of adjacent atoms or functional groups. To examine this effect, denote the molecules in Table 12 by RX–CF_3, where X is a bridging atom between the R group and the CF_3 group. Then the C–F distances in RX–CF_3 compounds correlate well with the magnitude of the electronegativity assigned to the X atom. This is shown in Fig. 5. The C–F bond lengths decrease approximately linearly with increasing electronegativity of X. At the same time note the large scatter of values when X is C. This suggest the operation of a higher order environmental factor which controls the effective electronegativity of X. Obviously when the R fragment incorporates highly electronegative atoms (F, O, N, etc.) the electron charge distribution on X is controlled by them. In such cases, the C–F distance in the CF_3 group accordingly decreases: for example, compare CF_3–CF_3, CF_3–CHO, $(CF_3)_3COH$, $(CF_3)_2C{=}NH$, etc., with CH_3–CF_3 and $CH_3{=}CH$–CF_3 (see Table 12). However, extended correlations based on higher order perturbations of X by R should be made with caution since the experimental data in Table 12 are of variable reliability.

An unexpected characteristic of the –CF_3 group is the correlation between the C–F distance and the FCF angle, as illustrated in Fig. 6. In spite of the considerable scatter in the data it is evident that the larger angles correlate with the shorter bond lengths. Note that most of the angles in Table 12 are based on r_g structures; these "angles" should be corrected (less than 1°) for the effects of molecular vibrations, when presenting a strict discussion of the geometrical structures. Were such corrections inserted, the above cor-

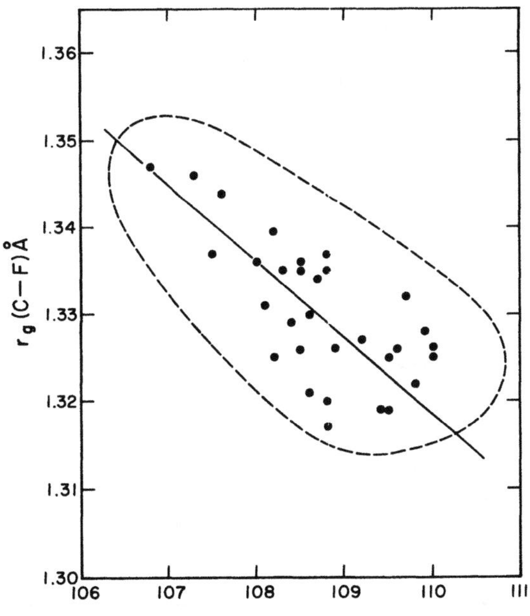

Fig. 6. C–F bond distances correlate with the FCF angle in –CF_3 groups––– values taken from Table 12. The closed curve represents an estimated error range based on the uncertainties of the values in Table 12

relations would be clearer since several of the data [$CH_2{=}CH{-}CF_3$[97] and $(CF_3)_3COH$[105]] for which r_α structures were given fall closer to the mean line.

In summary, the structures of CF_3 groups in RX–CF_3 compounds may be characterized as follows: (i) If X is a highly electronegative atom the C–F distance is shorter, relative to compounds with less electronegative X's. (ii) When the C–F distance is short, the FCF angle is large. (iii) Inductive effects on X due to the R group, produce a parallel trend. These can be simply rationalized:

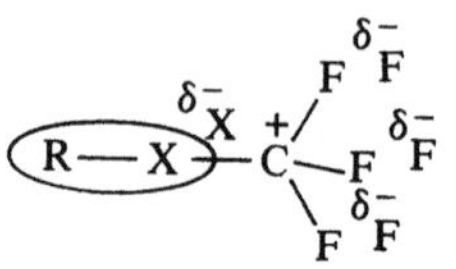

Because the electronegative F and X atoms accumulate relative negative charges (δ_F^- and δ_X^-), the central carbon assumes a positive charge which is greater the higher the electronegativity of X. Coulombic forces tend to shorten the C–F distance, while opening the FCF angle. These trends are also accounted for by the VSEPR model.[74]

II. D. Summary of Structural Data on C–F Bonds in Fluorocarbons

The above survey of C–F bond lengths and FCF angles in a variety of compounds was arranged on the basis of the number of fluorine atoms attached to the central atom. The full range of C–F distances observed is indicated in Fig. 7. The correlating parameters include not only the number of fluorine atoms attached to the central carbon, but also the state of hybridization assigned to the carbon atom and environmental factors controlled by the heteroatoms or functional groups adjacent to the carbon in question. To partially quantify these factors, assume that the C–F distance, R_{CF}, may be expressed as a function of the parameters F_n, C_h, and X_E, where F_n is the number of fluorine atoms, C_h is

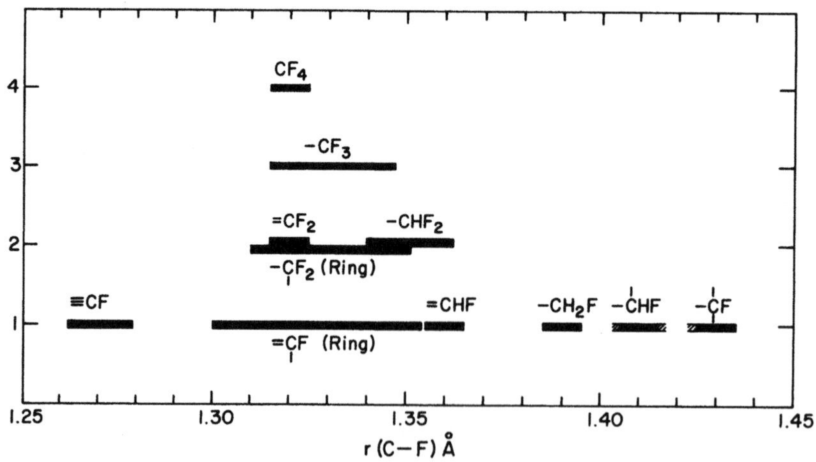

Fig. 7. Types of C–F bond distances arranged according to the numbers (n) of fluorine atoms attached to a given carbon atom.

a parameter which measures the carbon atom hybridization, and X_E represents the environmental factor. Then

(1) F_n: Typically in the fluoromethanes, $\Delta R_{CF} \approx -0.03$ Å for $\Delta F_n = 1$.
(2) C_h: Referred to sp^3, $\Delta R_{CF} \approx -0.04$ Å (sp^2) and -0.12 Å (sp). The environment hybridization factor for strained rings leads to $\Delta R_{CF} \approx |0.04|$ Å.
(3) X_E: (a) For primary, secondary, and tertiary carbon atoms in monofluorocarbons: $\Delta R_{CF} \approx |0.04|$ Å.
(b) In RCHF or RCF_2, where R is O, $=CF_2$ and $=CH_2$: $\Delta R_{CF} \approx |0.01|$ Å.
(c) In RCF, where R is ≡N and ≡CH: $\Delta R_{CF} \approx |0.02|$ Å.
(d) In fluorobenzenes: $\Delta R_{CF} \approx |0.05|$ Å.
(e) In RCF_3: $\Delta R_{CF} \approx |0.03|$ Å.
(f) In RCF_2 where R is H_2 and Cl_2: $\Delta R_{CF} \approx |0.02|$ Å.

The FCF angles in $>CF_2$, $=CF_2$, and $-CF_3$, are nearly constant (108 ± 2°), apparently independent of F_n, C_h, X_E, with the exceptions of perfluoroethylene and the three-member rings (in perfluorocyclopropane and perfluorodiazirine, ∠FCF = 112°).

III. Fluorine Substitution Effects

A.1. Change in Molecular Structure Adjacent to F Substitution Sites

In this section we shall examine the effect on the overall molecular structure of replacing hydrogen atoms by fluorines. Changes in the structural parameters are anticipated because the introduction of fluorine atoms substantially alters the electron distribution within the molecule. Hilderbrandt and Bauer[1,108)] called attention to such geometrical changes and classified them into two types:
Case 1 is represented by F_3C-X, $F_2C=X$, or FC≡X, where X is N, C, O, or a halogen, but not an atom which is double-bonded to an adjacent atom. Here the substitution of fluorine for hydrogen decreases the C–X bond length, with one notable exception, perfluoroethane, where the C–C distance is slightly larger ($r_g = 1.545 \pm 0.006$ Å),[113)] compared to that in ethane ($r_g = 1.533 \pm 0.002$ Å)[43)].
Case 2 is represented by $F_3C-X=Y$, where X is C or N, and Y is C, N, or O. Here the available data indicate a lengthening of the C–X bond.

Tokue, *et al.*[97)] recently investigated the structures of $CF_3-CH=CH_2$ and $CH_3-CH=CH_2$, by combining electron diffraction data with microwave spectra. For these the trend appears to be opposite to the generalization for *Case 2*. The need for a more critical analysis of molecular structures of fluorinated compounds is indicated, to establish more precisely the correlations between increments in structure and fluoro-substitution.

A.2. The Effects of Substitution of CH_3 by CF_3

The magnitudes of C–A distances in $(CH_3)_nA$ and $(CF_3)_nA$ (n = 1,2,3), where A is a specified central atom, are compared in Table 13. An interesting aspect is the difference $\Delta(r_F - r_H)$, introduced by CH_3/CF_3 substitution, $[\Delta(r_F - r_H) \equiv r(F_3C-A) - r(H_3C-A)]$. The C–A distance either decreases or increases depending on the atom A. Again, the correlation parameter is the electronegativity of A. From Fig. 8 it is evident that substitution

Table 13. C–A Distances in $(CH_3)_nA$ and $(CF_3)_nA$[1] (n = 1, 2, 3)

	A	r_F(C–A) Å	r_H(C–A) Å	$\Delta(r_F-r_H)$ Å
n = 1	F	r_a 1.320 ± 0.008[2]	r_a 1.391 ± 0.005[25]	−0.071
	Cl	r_a 1.751 ± 0.004[110]	r_a 1.784 ± 0.003[110]	−0.033
	Br	r_a 1.910 ± 0.006[99]	r_s 1.939[26]	(−0.029)
	I	r_a 2.115 ± 0.023[3]	r_s 2.139[26]	(−0.024)
n = 2	O	r_g 1.360 ± 0.005[121]	r_g 1.417 ± 0.003[122]	−0.057
	S	r_a 1.828 ± 0.015[123]	r_a 1.82 ± 0.01[8]	+0.008
			r_s 1.802 ± 0.002[124]	
	Se	r_a 1.978 ± 0.009[109]	r_a 1.977 ± 0.012[125]	+0.001
n = 3	N	r_a 1.43 ± 0.03[126]	r_a 1.455 ± 0.002[127]	−0.025
	P	r_g 1.939 ± 0.017[128]	r_g 1.847 ± 0.003[129]	+0.092
	As	r_a 2.05 ± 0.02[128]	r_a 1.98 ± 0.02[130]	+0.070
	(CH)	r_g 1.539 ± 0.003[35]	r_g 1.535 ± 0.001[45]	+0.004

[1] r_F(C–A) and r_H(C–A) represent the C–A distance in $(CF_3)_nA$ and $(CH_3)_nA$, respectively; $\Delta(r_F-r_H)$ is their difference. Note that for A=Br and I, r_a values are compared with r_s.
[2] Mean of values in Refs. [25, 120].
[3] Mean of values in Refs. [99, 111].

of CH_3 by CF_3 in $(CH_3)_nA$ makes the C–A distance shorter if the electronegativity of A is larger than 2.5 (that of carbon); conversely, C–A is longer when the electronegativity of A is smaller than 2.5.

For a wider class of compounds the C–X distances in CH_3X compared with CF_3–X, where X is a functional group or a fragment in a molecule, are listed in Table 14; $\Delta(r_F-r_H) \equiv r(F_3C-X) - r(H_3C-X)$. For most pairs there is a significant change in the C–X distance due to CH_3/CF_3 substitution. In particular, compounds $(CH_3)_2AA/(CF_3)_2AA$, where AA is –O–O–, –N=N–, –C≡C–, –S–S–, or –Se–Se–, follow a pattern similar to that illustrated in Fig. 8; the increment correlates with the electronegativity of the A atom. Another feature appears on close inspection of this table. When X is –COR, $-NO_2$, –NO, $-CF_3$, and –CN, the $r(F_3C-X)$ distance is longer than the $r(H_3C-X)$, suggesting a *destabilizing* interaction. Note that these functional groups follow the $-I^-$ substituents sequence as classified by Pople and Gordon.[141] For contrast, examine the $-I^+$ substituents, such as –F, $-NH_2$, –OR, etc. Unfortunately except when X is F no structural parameters for the $-I^+$ groups are available. By analogy with fluorine, one may anticipate for all $-I^+$ substituents there would be a shortening of the F_3C–X bond, compared to that for H_3C–X. When X represents hydrocarbon fragments ($-CH_3$, $-CH=CH_2$, –C≡CH, etc.), $r(F_3C-X)$ is generally shorter than $r(H_3C-X)$; this indicates a stabilizing interaction. In summary, the effect of CF_3 for CH_3 substitution on the magnitudes of the C–X bond length is:

(1) The C–X distance *increases* in the order:
X → $-NO_2$ > –NO > $-COCH_3$ > $-CF_3$ > –C≡N.

(2) The C–X distance *decreases* in the order:
X → –F > –Cl > (–Br) > (–I) > $-CH_3$ > $-CH=CH_2$ > –C≡CH.

Table 14. Comparison of C–X Distances in CF_3–X and CH_3–X, where X is a functional group or a fragment in a molecule.[1)]

X	$r_F(F_3C–X)$ Å	Refs.	$r_H(H_3C–X)$ Å	Refs.	$\Delta(r_F-r_H)$ $\times 10^{-3}$ Å
$-CH_3$	1.514 ± 0.014	95)	1.533 ± 0.002	43)	−19
$-CH{=}CH_2$	1.495 ± 0.006	97)	1.506 ± 0.003	97)	−11
$>C{=}CH_2$	1.535 ± 0.006	108)	1.505 ± 0.003	131)	+30
$-C{\equiv}CH$	1.465 ± 0.02	98)	1.470 ± 0.004	2)	−5
$-C{\equiv}CCH_3$	(1.464)	102)	1.467 ± 0.001	132)	(−3)
−CHO	1.540 ± 0.02	95)	1.514 ± 0.005	133)	+26
$-COCH_3$	1.562 ± 0.011	107)	1.517 ± 0.005	133, 108)	+45
$>CO$	1.551 ± 0.008	108)	1.517 ± 0.005	133, 108)	+34
$-CF_3$	1.545 ± 0.006	113)	1.514 ± 0.014	95)	+31
$-C{\equiv}N$	1.495 ± 0.014	99)	1.466 ± 0.003	2)	+29
−NO	1.546 ± 0.008	114)	(r_s) 1.490 ± 0.03	134)	(+56)
$-NO_2$	1.560 ± 0.02	116)	(r_s) 1.489 ± 0.005	135)	(+71)
$-N{=}NCH_3$	1.476 ± 0.005	101)	1.474 ± 0.001	101)	+2
$-O_2-$	1.398 ± 0.006	121)	1.445 ± 0.02	8)	−47
−N=N−	1.490 ± 0.002	101)	1.474 ± 0.001	101)	+12
$-C{\equiv}C-$	1.476 ± 0.005	2)	1.467 ± 0.001	132)	+9
$-S_2-$	1.829 ± 0.017	123)	1.806 ± 0.002	136)	+23
$-Se_2-$	2.018 ± 0.02	112)	1.954 ± 0.005	137)	+64
$-PH_2$	(r_s) 1.900 ± 0.006	138)	(r_s) 1.863 1.858 ± 0.003	139) 140)	+37

1) These are r_g distances, unless denoted otherwise; r_a values in the literature were converted to r_g's.
2) Mean of values (r_g) in Refs. 103, 104).

In the last category of effects due to CF_3/CH_3 substitution, consider the molecules with two or more CF_3 groups on one carbon atom: for example,

$$(CF_3)_2C{=}CH_2,\ (CF_3)_2C{=}O,\ (CF_3)_2C{=}NH,\ (CF_3)_3CH \text{ and } (CF_3)_3COH.$$

In all compounds investigated to date the C–C distance is longer in $(CF_3)_nCX$ than in $(CH_3)_nCX$. When X contains O or N the effect is larger than for carbon.

A.3. Substitution Effects in X–C(=O)–R Compounds

Egger and Cocks[142)] made the observation that RC(=O)–H and RC(=O)–alkyl bonds have lower dissociation energies than the corresponding bonds in the isoelectronic RC(=CH_2)–X. Structural data show a parallel trend; C–X bond distances in RC(=O)–X are longer than in RC(=CH_2)–X. This

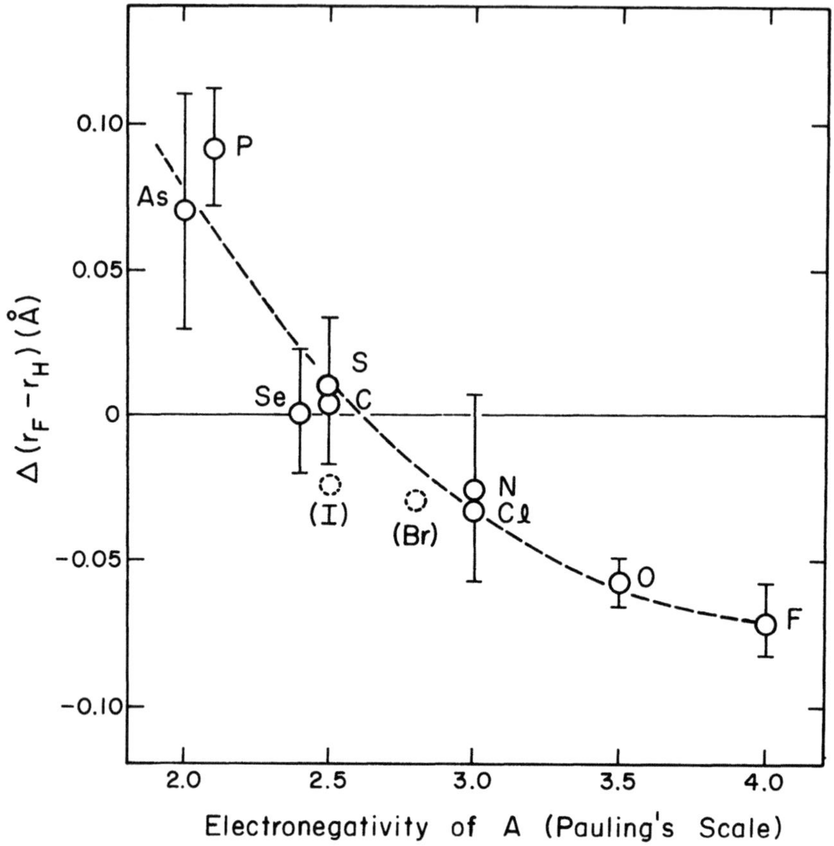

Fig. 8. Correlation between $\Delta\,(r_F - r_H)$, as listed in Table 13, and the electronegativity of A

is illustrated by the butadiene-acrolein-glyoxal series, and by the vinylacetylene-propynal series (Kuchitsu, *et al.*[2, 143]). They discussed the lenthening of C–X bonds in terms of a "heteroatom" effect (or, a secondary environmental effect due to the oxygen).

In the following tables we assembled data on the X–COR compounds and categorized C–X bond lengths in terms of an "interaction between function groups". (–X and –COR), as was done for the CF_3–X sequence. But instead of focusing attention on $\mathrm{R}\underset{\underset{\mathrm{O}}{\|}}{\mathrm{C}}\mathrm{-X}$ vs $\mathrm{R}\underset{\underset{\mathrm{CH_2}}{\|}}{\mathrm{C}}\mathrm{-X}$, we compared C–X distances in $\mathrm{R}\underset{\underset{\mathrm{O}}{\|}}{\mathrm{C}}\mathrm{-X}$ relative to H_3C–X. In Tables 15 and 16 C=O distances and OCX angles are also listed. As mentioned in Section II. A. 4, C(sp^2)–X bond distances are shorter than C(sp^3)–X distances. However, note that $\Delta[\equiv r(\mathrm{X{-}COR}) - r(\mathrm{X{-}CH_3})]$ assumes both positive and negative values depending on X; where X is $-NH_2$, –OH, –F, $-CH_3$, $-CH{=}CH_2$, and –C≡CH, Δ is negative; where X is – Br, $-CF_3$, –H, –Cl, –CHO, and –CN, Δ is positive. The magnitude of the *shorten-*

Table 15. Structures of X–CHO and X–CH_3[1])

X		r (C–X) Å	r (C=O) Å	∠ OCX	Refs.		r (H_3C–X) Å	Δ (× 10^{-3} Å)[2])
–NH_2	r_s	1.352 ± 0.012	1.219 ± 0.012	124.7° ± 0.3°	150)	r_o	1.474 ± 0.005[151)]	(–122)
–OH		1.361 ± 0.003	1.217 ± 0.003	123.4° ± 0.5°	146)		1.427[122)]	–66
–F	r_s	1.338 ± 0.005	1.181 ± 0.005	122.8° ± 0.5°	14)	r_s	1.385[26)]	–47
–CH=CH_2		1.484 ± 0.004	1.217 ± 0.003	124.0° ± 0.7°	144)		1.506 ± 0.003[97)]	–22
–CH_3		1.514 ± 0.005	1.209 ± 0.003	124.2° ± 0.5°	133)		1.533 ± 0.002[43)]	–19
–C≡CH		1.453 ± 0.003	1.214 ± 0.005	124.2° ± 0.2°	147)		1.470 ± 0.004[2)]	–17
–CHO		1.526 ± 0.003	1.212 ± 0.002	121.2° ± 0.2°	144)		1.514 ± 0.005[133)]	+12
–H		1.123	1.209 ± 0.003	119.2°	2, 143)		1.107 ± 0.001[149)]	+16
–CF_3		1.542 ± 0.02	1.201 ± 0.014	121.8° ± 4.5°	95)		1.514 ± 0.014[95)]	+28

1) The distances are r_g, unless otherwise indicated.

2) Δ = r (X–CHO) – r (X–CH_3).

Table 16. Structures of X–$COCH_3$ and X–CH_3[1])

X		r (C–X) Å	r (C=O) Å	∠ OCX	Refs.		r (H_3C–X) Å	Δ (× 10^{-3} Å)[2])
–NH_2		1.380 ± 0.004	1.220 ± 0.003	122.0° ± 0.6°	152)		1.467 ± 0.002[156)]	–87
–OH		1.365 ± 0.003	1.215 ± 0.003	122.8° ± 1.2°	153)		1.427[122)]	–62
–F	r_s	1.348	1.181	121.3°	55)	r_s	1.385[66)]	–37
–CH_3		1.509 ± 0.005	1.211 ± 0.003	121.7° ± 0.6°	108, 133)		1.533[43)]	–24
–CN		1.477 ± 0.008	1.208 ± 0.009	121.3°	154)		1.468 ± 0.002[66)]	+9
–Cl		1.798 ± 0.003	1.187 ± 0.003	120.3° ± 0.6°	155)		1.786 ± 0.003[110)]	+13
–H		1.127 ± 0.009	1.208 ± 0.004	120.5° ± 0.5°	2, 133)		1.107 ± 0.001[149)]	+20
–Br		1.975 ± 0.004	1.184 ± 0.004	120.8° ± 0.9°	155)	r_s	1.939[66)]	(+36)
–CF_3		1.562 ± 0.011	1.207 ± 0.006	116.8° ± 0.7°	107)		1.514 ± 0.014[95)]	+48

1) The distances are r_g, unless otherwise indicated.

2) Δ = r (X–$COCH_3$) – r (X–CH_3).

ing in $r(X-\overset{O}{\overset{\|}{C}}R)$ decreases in the order $-NH_2 > -OH > -F > -CH_3 \gtrsim -CH{=}CH_2 \gtrsim -C{\equiv}CH$.

The magnitude of the *lengthening* in $r(X-\overset{O}{\overset{\|}{C}}R)$ decreases in the order $-CF_3 > (-Br) > -H > -Cl, -CHO > -CN$. These sequences also correlate with the C=O distances in corresponding compounds, as illustrated in Fig. 9.

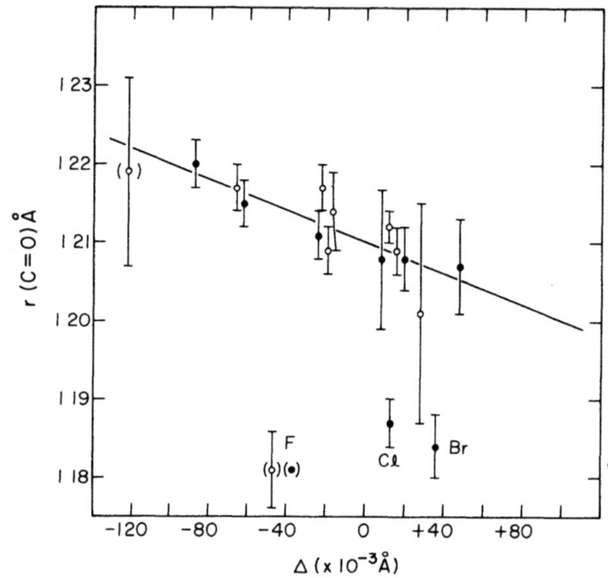

Fig. 9. Correlation between C=O distances and Δ as listed in Tables 15 and 16. The full circles show $X-COCH_3$ and the open circles are for X–CHO. The parentheses represent r_s structures, the others are r_g values. For proper comparison a few thousandths of an Å should be added to the r_s distances

The above sequence for bond length reduction is similar to that for π-electron delocalization in X–CH=Y molecules ($-CH_3$, –F, –OH, and $-NH_2$ for X, and $=CH_2$, =NH, and =O for Y) as calculated by Radom, Hehre and Pople[144], on the basis of a modest *ab initio* molecular orbital theory. Both π donation by X and π acceptance by Y increase for the sequence $CH_3 < F < OH < NH_2$. The bond separation energies, which measure interactions between groups (–X and –CH=Y), increase in the same order. Hence they suggested that the stabilizing interaction between the groups may be rationalized in terms of the π-electron transfer, $\overset{\frown}{X-C}H\overset{\frown}{=Y}$. Thus, for the four X's investigated by Pople, *et al.*[144] the sequence of C–X bond shortening in X–COR coincides with the ease of π-electron transfer. However, attention is called to the long C–C distance in $CH_3-CH{=}O$ (r_g = 1.514 ± 0.005 Å)[133], compared with the r_g(C–C) of 1.506 ± 0.003 Å[97] in $CH_3-CH{=}CH_2$; this is counter to Pople's model. The ease for π-electron transfer in $CH_3-CH{=}O$ is greater than for $CH_3-CH{=}CH_2$, and the C–C bond separation energy for the former is higher than for the latter, according to their calculations. In this connection, an attempt to account for the differences in the C–C distances in $CH_2{=}CH{-}C{\equiv}CH$, $CH_2{=}CH{-}C{\equiv}N$, and $OHC{-}C{\equiv}CH$ in terms of the π-bond order proved unsatisfactory (Fukuyama and Kuchitsu[70]). Indeed, it appears that the substituent effects on adjacent bond distances do not correlate with any single physical quantity. This is indicated by the *ab initio* calculations of C=C and C–F distances in $CH_2{=}CH_2$, $CH_2{=}CHF$, $CHF{=}CHF$, and $CH_2{=}CF_2$ by Bak, *et al.*[145]. Their results do show the observed trend, but neither charge densities due to π-overlap nor $(\sigma + \pi)$-overlap in the C=C and C–F bonds regions account for the computed bond distance changes; *i.e.* larger charge overlap did not always correspond to shorter distance, and *vice versa*.

Although the principal factors that determine variations in the bond length from molecule to molecule for a given atom pair are complex, one would anticipate C–X distances in X–C(=O)R to be shorter than in $X{-}CH_3$, simply due to the different states of hybridization of the carbon atoms. However, when X is $-CF_3$, –Br, –H, –Cl, –CHO, and –CN, the C–X distance in X–C(=O)R is longer. No explanation has yet been proposed.

It is instructive to compare $r(X{-}CF_3)$ and r(X–COR), for different types of X substituents. For convenience classify the X groups into three types:

S^+: $-NO_2$, –NO, –COR, $-CF_3$, –H, $-C{\equiv}N$
S^-: $-NH_2$, –OH, –F, $-CH_3$, $-CH{=}CH_2$, $-C{\equiv}CH$
$S^{\pm}$: Cl, Br, (I)

Referred to $r(X{-}CH_3)$, S^+ substituents increase both $r(X{-}CF_3)$ and r(X–COR) while S^- substituents shorten them; some combinations ($CF_3{-}OH$, $CF_3{-}NH_2$, $O_2N{-}COR$ and ON–COR) have not yet been investigated. In contrast, $S^{\pm}$ substituents exert opposite effects on the $r(X{-}CF_3)$ and r(X–COR); the former is shortened and the latter is lengthened. The structure of $I{-}COCH_3$ has been reported by S. Tsuchiya [J. Mol. Struct. *22*, 77 (1974)]: r_g (C–I) = 2.217 ± 0.009 Å, which is about 0.07 Å longer than r_s (C–I) in CH_3I (Ref.[8]).

In summary, r(C=O) distances and ∠OCX angles in X–COR compounds (Tables 15 and 16) show the following trends:

(1) The ∠OCX angle of the S^+ type substituents (121°–125°) tend to be larger than that for S^- types (117°–122°).
(2) Except for the halogens (F, Cl, and Br) the C=O distances in X–COR correlate with $\Delta\,[= r(X{-}COR) - r(X{-}CH_3)]$; when r(X–COR) is short, r(C=O) is long, and *vice versa.* (See Fig. 9).
(3) When X is a halogen r(C=O) is about 1.185 Å, which is shorter than any other r(C=O) in X–COR.

A.4. C–C, C=C, and C≡C Distances in Fluorocarbons

It is generally accepted that carbon-carbon bond lengths (C–C, C=C, and C≡C) in the fluorocarbons are shorter than in the corresponding hydrocarbons. Typical examples are

Table 17. The effect of monohalogenation on the C–C distance in alkanes, (Å units)

X	Primary CH_3CH_2X	Δ[1]	Secondary $(CH_3)_2CHX$	Δ[1]	Tertiary $(CH_3)_3CX$	Δ[1]
H	r_g 1.532$_6$ ± 0.002[43]		r_g 1.532$_3$ ± 0.003[157]	0	r_g 1.535 ± 0.001[45]	+2
	r_s 1.526[44]	–	r_s 1.526 ± 0.002[42]	0	r_s 1.525 ± 0.001[44]	–1
F	r_s 1.505 ± 0.004[26]	–21	r_g 1.517 ± 0.002[34]	–15	r_g 1.522 ± 0.008[158]	–11
Cl	r_s 1.520 ± 0.003[37]	–6	r_s 1.522 $^{+0.004}_{-0.006}$[38]	–4	r_s 1.530 ± 0.002[39]	+4
					r_g 1.528 ± 0.002[48]	–5

[1] C–C bond distance referred to that in ethane (× 10^{-3} Å unit). The distances are compared for the same measure (r_g or r_s).

Table 18. The C–C distances in fluorinated ethanes

Compounds	r(C–C) Å	Δ (× 10^{-3} Å)[1]
CH_3–CH_2F	(r_s) 1.505 ± 0.004[27]	–21
CH_3–CHF_2	–	(?)
CH_3–CF_3	(r_g) 1.514 ± 0.014[95]	–19
	(r_o) 1.492[2]	
	(r_o) 1.530[96]	
FCH_2–CH_2F	(r_g) 1.538 ± 0.002[3]	+5
F_2CH–CHF_2	–	(?)
F_3C–CF_3	(r_g) 1.545 ± 0.006[113]	+12
	(r_a) 1.560 ± 0.03[158]	
F_3C–CH_2F	–	(?)
F_3C–CHF_2	(r_s) 1.520[106]	–6

[1] The difference in the C–C distance, referred to (r_g) 1.533 Å and (r_s) 1.526 Å in ethane (Table 17).
[2] Thomas, L. F., Heeks, J. S., and Sheridan, J.: Z. Electrochem., *61*, 935 (1957).
[3] Estimated from the r_a distance in Ref.[32].

Table 19. The effect of X as a substituent on the length of the double bond in X–CH=CH_2

X	r_g (C=C) Å	Δ (× 10^{-3} Å)[1]	∠X–C=C	Refs.
–H	1.337 ± 0.002	–	121.9° ± 0.4°	161)
–F	1.333 ± 0.007	–4	121.0° ± 0.2°	50)
$-CF_3$	1.318 ± 0.008	–19	125.8° ± 1.1°	97)
$-CH_3$	1.342 ± 0.002	+5	123.4° ± 0.4°	97)
–Cl	1.355 ± 0.008	+18	121.1° ± 0.9°	162)
–CHO	1.345 ± 0.003	+8	121.0° ± 0.7°	143)
–C≡N	1.343 ± 0.004	+6	121.7° ± 0.5°	70)
–C≡CH	1.344 ± 0.004	+7	123.1° ± 0.5°	69)
–CH=CH_2	1.345 ± 0.002	+8	123.6° ± 0.3°	143)

[1] Differences of r_g (C=C), referred to r_g (H_2C=CH_2).

Table 20. C=C Distances in Substituted Ethylenes

Compound	r_g (C=C) Å[1]	Δ (x 10^{-3} Å)[2]	Refs.
$H_2C{=}CH_2$	1.337 ± 0.002	–	161)
$F_2C{=}CH_2$	1.316 ± 0.006	−21	50)
FHC=CHF (*cis*)	1.331 ± 0.004	−6	50)
FHC=CHF (*trans*)	1.329 ± 0.004	−8	50)
$FHC{=}CF_2$	1.309 ± 0.006	−28	50)
$F_2C{=}CF_2$	1.311 ± 0.007	−26	50)
$(CF_3)_2C{=}CH_2$	1.374 ± 0.013	+37	108)
$(CH_3)_2C{=}CH_2$	1.342 ± 0.003	+5	2, 131)
BrHC=CHBr (*cis*)	1.360 ± 0.01	+23	163)
ClHC=CHCl (*cis*)	1.354 ± 0.005	+17	164)
$Cl_2C{=}CCl_2$	1.355 ± 0.003	+18	165)
$Br_2C{=}CBr_2$	1.363 ± 0.009	+26	166)
$I_2C{=}CI_2$	1.364 ± 0.015	+27	165)
$(CH_3)_2C{=}C(CH_3)_2$	1.351 ± 0.003	+14	167, 168)
$(CN)_2C{=}C(CN)_2$	1.358 ± 0.01	+21	169)
$(CH_3)CH{=}CH(CH_3)$ (*cis*)	1.347 ± 0.003	+10	170)
$(CH_3)CH{=}CH(CH_3)$ (*trans*)	1.348 ± 0.003	+11	170)

[1]) r_a values in the literature were converted to r_g values.
[2]) The differences are referred to the C=C distance in ethylene.

Table 21. The substitution effect of X on triple bonds in X–C≡CH and X–C≡N

X	r (C≡C) Å	Refs.	Δ[1])	r (C≡N) Å	Refs.	Δ[1])
–H	r_o 1.2088 ± 0.002	67)		r_o 1.1558	68)	
				r_s 1.1553 ± 0.001	26)	
	r_e 1.2036	67)		r_e 1.1538 ± 0.001	26)	
	r_g 1.212 ± 0.002	132)		r_g 1.157 ± 0.002	2, 66)	
–F	r_s 1.198	65)	−11	r_s 1.159	65)	+4
$-CH_3$	r_g 1.210 ± 0.004	2, 66)	−2	r_g 1.159 ± 0.002	2, 66)	+2
$-CF_3$	r_g 1.202 ± 0.004	98)	−10	r_g 1.151 ± 0.01	99)	−6
				r_o 1.153	100)	−3
–Cl	r_s 1.204	65)	−5	r_s 1.159	65)	+4
–Br	–			r_s 1.158	65)	+3
–I	–			r_s 1.159	65)	+4
–C≡N	r_s 1.205	65)	−4	r_g 1.162 ± 0.002	171, 172)	+5
–C≡CH	r_g 1.218 ± 0.002	172)	+6	r_s 1.159	65)	+4
$-CH{=}CH_2$	r_g 1.215 ± 0.003	69)	+3	r_g 1.167 ± 0.004	70)	+10
$-CCl_3$	–			r_g 1.165	173)	+8
–CHO	r_g 1.211 ± 0.006	147)	−1	–		

[1]) Δ = [r (XC≡CH) – r (HC≡CH)] or [r (XC≡N) – r (HC≡N)]. (x 10^{-3} Å unit). In the former, the r_s distances are compared with the corresponding r_o distance, while the latter Δ's are compared according to the same measure of distance.

listed in Tables 17–21. The magnitude of the increment in the C–C distances is linearly related to the extent of alkane substitution (primary, secondary, and tertiary); see Fig. 10.

The C–C distances in poly-fluorinated ethanes provide a particularly interesting sequence, for which unfortunately key values are still missing (Table 18). It appears that α-fluorination leads to a shorter C–C distance, while α, β-fluorinations lengthen the C–C distance, with the exception of CF_3–CHF_2. Further studies of fluorinated ethanes are needed to provide data on the fluorine substitution effect.

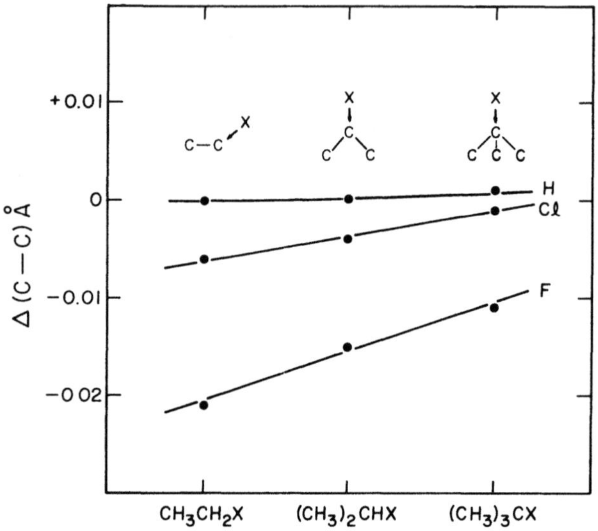

Fig. 10. Increments in C–C distances, referred to that in ethane, as present in primary, secondary, and tertiary compounds. Values for Δ (C–C) and the range of errors are listed in Table 17

Tables 19 and 20 show the effect of substitution on the C=C bond distance in ethylene derivatives. The striking feature is that all C=C distances except for the fluorinated compounds are significantly longer than that in ethylene; with the one notable exception, $(CF_3)_2C{=}CH_2$, all C=C distances in the fluorinated compounds are shorter. Attention is called to the pair $CF_3CH{=}CH_2$, $(CF_3)_2C{=}CH_2$; in the former both of the C–C and C=C distances are short while in the latter both of the C–C and C=C distances are long, compared to their hydrogen analogs (Tables 14, 19, and 20). In contrast to the fluoroethanes, in both α- and α,β-fluorinated ethylenes, the C=C distances are short; the effect is particularly marked for $F_2C{=}CH_2$, $F_2C{=}CHF$, and $F_2C{=}CF_2$. While no physically interpretable explanation for these C=C contractions due to the fluorination have been proposed, *ab initio* MO calculations of $CH_2{=}CH_2$, $CH_2{=}CHF$, cis-CHF=CHF, and $CH_2{=}CF_2$ by Bak, *et al.*[145] do show trends similar to the observed. Pappas[160] considered the influence of electronegative substituents on hybridization, and tried to explain the shortening of the C=C bond in $CH_2{=}CF_2$ in terms of a *re*hybridization on the carbon atom by the electronegative fluorine. His estimated shortening relative to CH_2CH_2 of 0.012 Å is about half of the observed shortening, 0.021 ± 0.008 Å.

Table 21 shows the effect on triple bonds due to the X substituent in X–C≡N. The C≡C bonds, relative to HC≡CH, are hardly changed except when X is –F and –CF_3; then the increment is ≈−0.01 Å. When X is −Cl, and −CN, the C≡C distances are slightly shortened (≈0.005 Å) whereas when X is $-CH_3$, $-CH{=}CH_2$ −CHO, and −C≡CH the C≡C distances remain unchanged (± 0.003 Å), or become slightly longer (0.006 Å when X is −C≡CH). α,β-substitution on −C≡C− by CF_3 groups, $CF_3{-}C{\equiv}C{-}CF_3$, leads to a shorter C≡C distance [r_g = 1.199 ± 0.003 Å[103] or r_a = 1.204 ± 0.003 Å[104]] by about 0.01 Å, compared to those in $CH_3{-}C{\equiv}C{-}CH_3$ [r_g = 1.213 ± 0.001 Å[132]] and HC≡CH. On the other hand, all the C≡N distances in X−C≡N appear to be lengthened, except when X is $-CF_3$; here the magnitude of the increment relative to HCN is within the experimental error.

From the above compilation of structural data for the fluorinated hydrocarbons one may summarize the differences between the geometrical structures of the fluorinated vs the parent species, as follows:

(α) When one or more fluorine atoms replace hydrogens on a single carbon atom, irrespective of the connectivities, the adjacent C−C distances are decreased.

(β) When fluorines replace hydrogens on several neighboring carbons, the net effect depends on the bond types within the carbon skeleton. Adjacent single bonds are lengthened, whereas multiple bonds are generally shorter than in the parent hydrocarbons.

A.5. Change of Ring Size Upon F for H Replacement

Ring dimensions of small perfluorocyclic compounds are compared with their hydrocarbon analogues in Table 22. The perfluorocyclopropane ring is smaller (ca. −0.007 Å), whereas the ring dimensions in perfluorocyclobutane and cyclohexane are larger by about 0.009 Å and 0.016 Å, respectively. This is reminescent of the situation for perfluoroethylene and perfluoroethane; the C=C bond in the former is smaller while the C−C distance in the latter is larger than in ethylene and ethane, respectively. This parallelism underscores the observation that cyclopropane is an ethylene-like molecule.

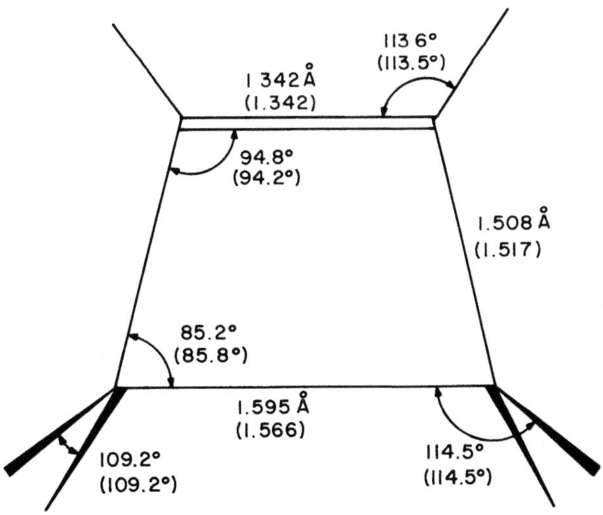

Fig. 11. Structures of cyclobutene[175] (r_s values are in parentheses) and perfluorocyclobutene (r_g)[77]

Table 22. Comparison of ring dimensions in perfluorocarbons with their hydrogen analogs (Å)[1])

Ring		H	F
		1.512 ± 0.003[79])	1.505 ± 0.003[76])
		1.557 ± 0.003[2]) 1.550 ± 0.003[80])	1.566 ± 0.008[77])
		1.546 ± 0.001[81])	–
		1.536 ± 0.002[82])	1.552 ± 0.001[78])
	CN NN	1.482 ± 0.003[174]) (r_s) 1.228 ± 0.003	1.428 ± 0.004[92]) 1.294 ± 0.009
	C–C [2]) C=C	1.533 ± 0.005[175]) (r_s) 1.342 ± 0.005	1.537 ± 0.003[77]) 1.342 ± 0.006
	C–C [2]) C=C	1.533 ± 0.007[176]) 1.343 ± 0.01	1.525[93]) 1.342 (assumed)
		1.399 ± 0.002[34]) 1.400[177])	1.396 ± 0.007[63]) 1.408 ± 0.006[60]) (r_o) 1.397[61])

[1]) r_a distances in the literature were converted into the r_g, unless noted otherwise.
[2]) Weighted averages of the C–C single bonds.

The ring dimensions of diazirine and perfluorodiazirine differ significantly.[92, 174]) In F_2CN_2 the C–N distance is shorter than in H_2CN_2 by about 0.05 Å; whereas the N=N bond length is longer by about 0.06 Å. Although the comparison between the two compounds was made according to different measures (r_s and r_g), the indicated differences are striking. Note that extended Hückel MO calculations of electron overlap populations in H_2CN_2 and F_2CN_2 show less density for N=N in the perfluoro compound than in diazirine and inversely for the C–N bonds. These correlate with the observed relative distances (Hencher and Bauer[174])).

In perfluorocyclobutene, perfluorocyclopentene and perfluorobenzene the ring dimensions are essentially the same as in the corresponding hydrogen analogs. For the former two compounds, the weighted average values of the non-equivalent C–C single bonds are compared in Table 22. A detailed comparison is presented in Fig. 11, for cyclobutene (r_s)[175]) and perfluorocyclobutene (r_g)[77]). One may conclude that perfluorination does not alter the essential features of these unsaturated rings, keeping in mind that different measures of distance were used in specifying these structures. Furthermore, it should be noted that precise discussions of molecules such as perfluorocyclobutene, which incorporates several slightly unequal bond distances (C=C and C–F; =C–C and C–C) are indeed difficult on the basis of any diffraction technique, where correlations between such parameters and/or their mean amplitudes of vibration introduce ambiguities in the least squares analysis for the resolution of the small differences. Thus, in perfluorocyclopentene the attempt to resolve the two non-equivalent C–C bonds led to indistinguishable bond lengths but with large uncertainties (~0.04 Å), in spite of the imposed constraints on the magnitude of the C=C distance and several mean amplitudes[93]. This difficulty exists for the partially fluorinated benzenes with reference to possible distortions of the benzene rings due to unsymmetric substitution. For perfluorobenzene, however, the structure is well determined because of its high symmetry. In C_6F_6 the carbon ring size is essentially equal to that in benzene: 1.396 ± 0.007 Å[63] 1.408 ± 0.006 Å[60] vs 1.399 ± 0.002 Å.[34] The corresponding magnitudes in perchlorobenzene and perbromobenzene are 1.404 ± 0.002 Å[178] and 1.403 ± 0.005 Å (r_g),[166] respectively.

B.1. Conformations – General Remarks

In this section we are concerned with the effects of F for H substitution on the relative stabilities and molecular dynamic aspects of rotational isomers. There are three types of problems which require solution. To determine –

(a) the relative populations of conformers (*e.g. trans, gauche,* and/or *cis*) as a function of sample temperature;
(b) the potential barrier height hindering rotation around single bonds, or structural inversions; and
(c) the origins of such barriers.

The electron diffraction technique is one of several tools for investigating (a). In turn, such information, if sufficiently detailed, can provide estimates of barrier heights. Furthermore, thermally averaged mean amplitudes for non-bonded atom pairs, especially those engaged in large amplitudes of relative motion, can be obtained from electron diffraction data. These magnitudes are closely related to the potential functions which control such motions, and thus comprise a measure of intramolecular forces. Of the many precautions which one must observe in order to derive meaningful dynamical parameters, particular attention is called to the following two: (i) Use of correct sample temperatures. It has been generally assumed that the sample temperature is equal to the nozzle temperature. Depending on experimental design, due to jet expansion the translational and vibrational temperatures are $(0.7\text{–}0.8)T_{\text{nozzle}}$[179]. (ii) Adequate use of spectroscopic data, when they are available; *e.g.* vibrational force fields provide good estimates for many of the mean amplitudes which must be constrained during a least squares calculation of those mean amplitudes which are associated with the low frequencies.

The most accurate information on the magnitudes of potential barriers (b) is derived from microwave spectra. Review papers on the determination of barrier heights hindering internal rotation have been published by Wilson[180] and Lowe[181]. Extensive data on barriers are given in these summaries; see also Refs.[182–185].

As for (c), despite the many theoretical studies published to date no simple formulation, nor generally applicable theory has been presented. Lowe[186] classified recent theoretical analyses of barrier heights into three types. For ethane a simple and intuitively useful explanation for the origin of the barrier is possible[186]. In the case of molecules with lower symmetry or more polar bonds or both, additional factors must be introduced to provide an adequate description[187]. This is the case for the fluorocarbons reviewed in the next section.

B.2. Fluoroethanes and Related Compounds

It is both challenging and puzzling to attempt to correlate in the fluoroethanes the reported magnitudes of the three-fold barriers, V_3, for hindered rotation:
CH_3CH_3[188], $CH_3{-}CH_2F$[189], $CH_3{-}CHF_2$[189], $CH_3{-}CF_3$[190], $CF_3{-}CH_2F$[191], $CF_3{-}CHF_2$[10(
and $CF_3{-}CF_3$[113], (Fig. 12).

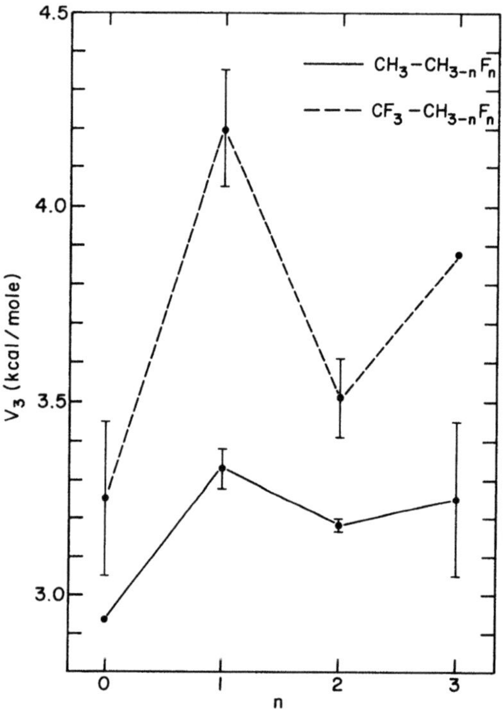

Fig. 12. Potential barrier heights for rotation about C–C bonds in fluorosubstituted ethanes

At either end the first F for H replacement raises the barrier height; the second fluorine on the same carbon ($CH_3{-}CHF_2$ or $CF_3{-}CHF_2$) lowers the barrier, and the final fluorination ($CH_3{-}CF_3$ or $CF_3{-}CF_3$), raises the barrier again. This alternation has been rationalized by Lowe and Parr[192], on the basis of an electrostatic model and the Hellmann-Feynman theorem. Their semi-empirical method has also been successfully applied to changes in barrier height in the fluorosubstituted methyl silanes[192], and to the replacement of H by various halogens in the secondary position in propene and in acetaldehyde[193]:

V_3 (cal/mole) = 1995 ($CH_3{-}CH{=}CH_2$)[194],
2440 ($CH_3{-}CF{=}CH_2$)[195],
2671 ($CH_3{-}CCl{=}CH_2$)[196],

1160 (CH_3–CHO)[197],
1040 (CH_3–CFO)[55],
1296 (CH_3–CClO)[198], and
1305 (CH_3–CBrO)[199]; refer also to Table 25.

Although the method is based upon a number of unproved assumptions, this approach has been more successful than the purely classical calculations, examplified by the Westheimer-Hendrickson procedure[200]. Barrier heights derived *via* the latter approach by Abraham and Parry[201], and Scott and Scheraga [202] did not reproduce the zig-zag trends found for the fluoroethanes. CNDO predictions for these barriers also failed in this respect.[203]

The 1,2-disubstituted ethanes have two rotational isomers (*gauche* and *trans*). Consideration of their relative stabilities provide test data for theories on intramolecular forces (*i.e.* non-bonded repulsion or attractions, intramolecular hydrogen bonding, etc.). Structural and energetic data on the rotational isomers of XCH_2–CH_2Y species (where X, Y are F, Cl, Br, OH, NH_2, CN, or CH_3) in the gas phase are given in Table 23. For the combination

(X, Y) = (F, F), (F, CH_3), (Cl, CH_3), (OH, OH), (F, OH), (Cl, OH), and (NH_2, NH_2),

the most stable form is *gauche*. In contrast, the most stable form for the combinations

(X, Y) = (Cl, Cl), (Br, Br), (Cl, F), (Cl, Br), (CN, CN), and (CH_3, CH_3) is *trans*.

The energy differences between the *gauche* and *trans* forms are small, generally being less than 1~2 kcal/mole. The thermal average values for the dihedral angles in the *gauche* isomers are larger than 60°. However, one should recall that torsional angles derived from electron diffraction data incorporate large uncertainties, at least ~5°, since their magnitudes are strongly correlated with the amplitudes of torsional motion and sensitively depend on the procedure used for data reduction.

Concerning the conformational preferences of XCH_2–CH_2Y type molecules (where X, Y are F, NH_2, OH, or CH_3), Radom, *et al.* [218] examined factors which influence conformational preferences, using *ab initio* (LCAO-SCF) molecular orbital calculations. For the *gauche* conformation "steric" interactions are destabilizing while intramolecular hydrogen bonding is stabilizing. In the absence of the latter the interaction of two vicinal electron-withdrawing polar bonds is apparently destabilizing. Therefore, when both X and Y are either bulky or electronegative groups, the more stable conformer is *trans*, in the absence of hydrogen bonding. However, these explanations failed for 1,2-difluoroethane for which *trans* was predicted to have the lowest energy (1.0 or 1.47 kcal/mole more stable than the *gauche*), contrary to the observed *gauche*/*trans* ratio. For the other molecules Radom's analysis agreed with the experimental trends. Empirical calculations of the Westheimer-Hendrickson type by Abraham and Parry[201] and Balthuis, *et al.*[219] did not predict the *gauche* preference for 1,2-$C_2H_4F_2$, although results in agreement with observation were derived for the other 1,2-haloethanes. On the basis of CNDO calculations, Gordon[203] predicted that the *gauche* isomer of 1,2-difluoroethane would be more stable by 0.1 kcal/mole; but this agreement must be regarded as fortuitous, in light of the unreasonable results he obtained for the other fluoroethanes. Extended Hückel calculations by Pachler and Tollenaere[220] correctly indicated that in 1,2-difluoroethane the predominant form is *gauche*. However, their caculated energy differences between the *gauche* and *trans* isomers was too large; the same calculations led to a wrong barrier for monofluoroethane (*i.e.* lower than that for ethane; see Fig. 12).

Although the energy differences (E_g–E_t) for 1,2-difluoroethane obtained by the electron diffraction method[31,32] (Table 23) must be regarded as a rough estimate and little credence should be placed on the quantitative magnitudes, the sign of (E_g–E_t) appears to be correct. To deduce from diffraction data accurate energy differences between rotational isomers one must (i) compute the partition function of the isomers; (ii) estimate

Table 23. Conformations of $XCH_2{-}CH_2Y$ species with respect to the C–C axis (gas phase)

X	Y	Gauche (%)[5]		Φ[1]	$E_g{-}E_t$[2]	Method[3]
F	F	85 ~ 96	(14 °C)[4]	69°	−(0.6 ~ 1.4)	ED[31]
		~ 98	(Room)	74°	−(~ 1.7)	ED[32]
					0 (± 0.2)	IR[204]
Cl	Cl	27 ± 5	(22 °C)	71°		ED[205]
		27 ± 7	(−7 °C)	64°	+(0.8 ~ 1.0)	ED[31]
		~ 25	(83 °C)		+(1.2 ± 0.1)	ED, IR, R and DM[206]
Br	Br	7 ~ 18	(22 °C)	74°	+(1.3 ~ 1.9)	ED[31]
					+(1.8 ± 0.2)	IR[182]
OH	OH	~ 100	(Room)	74°		ED[207]
NH_2	NH_2	95 ~ 100	(55–118 °C)	64°		ED[208]
CH_3	CH_3	~ 40	(Room)	65°	+(~ 0.65)	ED[209]
CN	CN				+(~ 1.0)	IR[210]
CH_3	F			63°	−(0.5 ± 0.3)	MW[211]
CH_3	Cl	81 ± 5	(Room)	59°	−(0.05 ~ 0.30)	ED[212]
OH	F	~ 100	(Room)	60 ~ 70°		ED[213]
		> 95	(156 °C)	65°	−(> 2.8)	ED[30]
OH	Cl	90 ~ 100	(37 °C)	61°	−(1.8 ~ 4.5)	ED[214]
		75 ~ 85	(200 °C)	70°	−(1.8 ~ 4.5)	ED[214]
					−(0.95 ± 0.02)	IR and R[215]
Cl	F				+(0.5)	DM[216]
Cl	Br				+(1.3 ± 0.1)	IR[217]

1) The dihedral angle (thermal average between the XCC and CCY planes) for the *gauche* form, measured from the *cis* position.
2) The energy difference between the *gauche* and *trans* forms (kcal/mole).
3) Experimental methods, ED: electron diffraction, IR: Infrared, R: Raman, MW: microwave, DM: dipole moment.
4) Reservoir temperature.
5) Remander is the *trans* isomer.

the temperature of the sample at the point of diffraction; and (iii) accurately measure the temperature dependence of the total diffraction pattern. Klaboe and Nielsen[204] measured the intensities of absorption for the strong infrared bands assigned as fundamentals for the two isomers of $CH_2F{-}CH_2F$ at 25°, 70°, 110°, 140°, and 170 °C. No change in the integrated extinction of any band could be detected, but the bands became broader at higher temperatures. Therefore, they concluded that the enthalpies of the two isomers in the gaseous state must be equal within the experimental error of about 0.2 kcal/mole. Their infrared and Raman studies[204] and NMR studies by Abraham and Kemp[221] suggest that the *gauche* conformation was preferred in the liquid state. It is not unusual that relative conformational stabilities differ for the gaseous, liquid and solid states due to neighbor-neighbor interactions.

B.3. Comments on Individual Molecules

1,2-Dihaloethylenes

The 1,2-disubstituted ethylenes, XHC=CHY, exist in two forms with the *cis* often more stable than the *trans*. For example, the equilibrium composition of FHC=CHF at 200° is

63% (*cis*) and 37% (*trans*).[222] Similarly for

(X, Y) = (F, Cl)[222], (F, Br)[222], (F, I)[223], (Cl, Cl)[224], (Br, Br)[225] (CH_3, Cl)[226] and (CH_3, Br)[227],

the *cis* fractions at equilibrium at corresponding temperatures are 70, 70, 67, 61, 50, 76, and 68%, respectively. The relative stabilities of the *cis* isomer relative to *trans* cannot be rationalized simply on the basis of steric effects, which imply repulsive interactions between nonbonded atoms or groups. Energetically favorable interaction between X, Y and the vicinal atoms or groups must be considered. Recently, the importance of "attractive nonbonded interactions" has been proposed.[228–230] On the basis of molecular orbital arguments Epiotis[64] suggested that the interaction between lone pairs or between unsaturated bonds may be attractive; also, that the symmetry of the molecule determines in a major way whether non-bonded interactions are attractive or repulsive. In the case of the *cis*- and *trans*-1,2-dihaloethylenes, the enhanced stability of the *cis* over the *trans* isomer can be explained by the net attraction between the two halogen atoms due to pπ interaction in the *cis* form. If the two fluorine atoms in *cis*-1,2-difluoroethylene do *attract* each other, one may anticipate an accompanying effect on the geometrical parameters: *i.e.*, smaller C=C–F bond angles or shorter C–F bonds in the *cis* than in the *trans* isomer. The structures of *cis* and *trans*-1,2-difluoroethylenes are illustrated in Fig. 13[50]. While the C=C bond bond distances in both forms are equal within the experimental errors, the C–F bonds and the C=C–F bond angles in the *cis* form are about 0.01 Å longer and 4.4° larger than those in the *trans* form, contrary to the above expectations. The structural evidence indicates that the two fluorine atoms in the *cis* form *repel* each other.

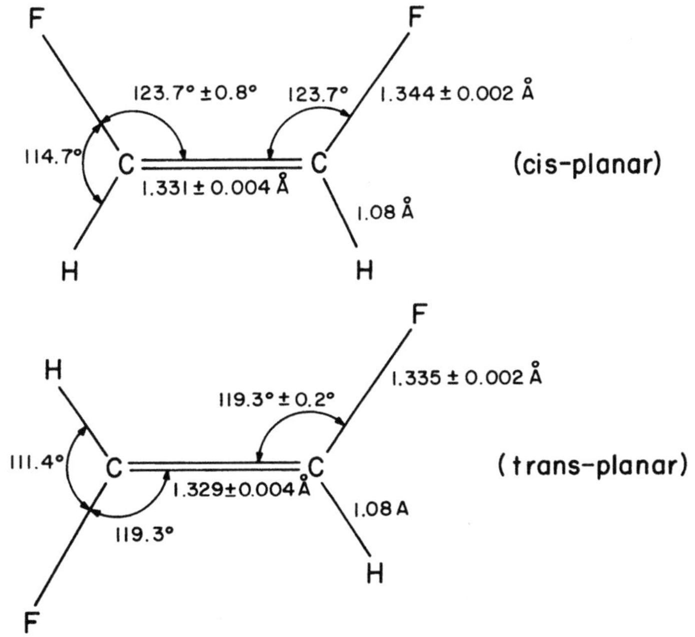

Fig. 13. Structural parameters (r_g) in *cis*- and *trans*-1,2-difluoroethylene.[50]

The Azocompounds: XN=NY

Bond distances and angles for XN=NY molecules (where X, Y are H, F, CH_3 or CF_3) are

Table 24. Bond distances and angles in XN=NY compounds

X	Y		r (N=N)Å	∠N=N–X ∠N=N–Y	Observed stable form	Refs.
H	H		1.252 ± 0.002	106.8° ± 0.5°	*trans*	231)
CH_3	D	(r_a)	1.245 ± 0.002	112.3° ± 0.4° 110.3° ± 5.1°	*trans*	232)
CH_3	CH_3	(r_g)	1.254 ± 0.002	111.9° ± 0.2°	*trans*	101, 233)
F	F	(r_g)	1.231 ± 0.01	105.5° ± 0.7°	*trans*	234)
F	F	(r_g)	1.214 ± 0.012	114.4° ± 1.0°	*cis*	234, 235)
CH_3	CF_3	(r_g)	1.219 ± 0.003	110.5° ± 1.3° 126.2° ± 0.6°	*trans*	101)
CF_3	CF_3	(r_g)	1.236 ± 0.005	133.0° ± 0.3°	(*cis*)[1])	101)

[1]) See text.

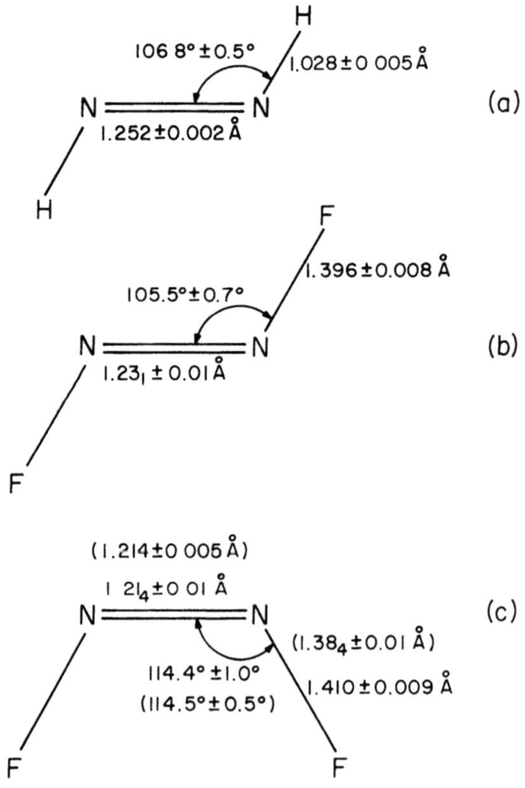

Fig. 14. Distances and bond angles in *trans* HN=NH, and in *trans* and *cis*-FN=NF. (*a*) Infrared data[231]); (*b*) Electron diffraction results[234]); (*c*) Electron diffraction values[234]) compared with microwave values[235]) (in parentheses)

listed in Table 24.[101, 231–235]) These are isoelectronic with XHC=CHY and may exist in two isomeric forms. However, in this series the existence of both *cis* and *trans* isomers has been experimentally confirmed only for FN=NF[234]) and azobenzene.[236]) The enthalpy of formation of *cis*-N_2F_2 is 3.0 kcal/mole lower than that for *trans*[237], indicating some stabilizing interaction between the vicinal fluorine atoms in the *cis* isomer, as proposed for *cis*-1,2-difluoroethylene. The geometrical parameters in N_2H_2 and N_2F_2 are illustrated in Fig. 14. Here the relative stability of the *cis* isomer is indicated by the shorter N=N bond length which also correlates with relative overlap population for the region between the nitrogen atoms.[238] On the other hand, the larger (≈9°) N=N–F angle in *cis* relative to trans *trans*-N_2F_2 follows the trend observed for the 1,2-difluoroethylene.

A rather puzzling structure in this series is hexafluoroazomethane which was reported to be in the *cis* form[101] whereas azomethane and 1,1,1-trifluoroazomethane have been confirmed to be the *trans* conformation only;[101] see Table 24. The relative stability of the *cis* configuration for $CF_3N{=}NCF_3$ was supported by an extended Hückel calculation of the energy as a function of the N=N–C angle, holding all other parameters constant.[101] These indicated that the *cis* isomer was more stable, with a minimum at 130° for the N=N–C angle. However, Chang, *et al.*[101] mentioned that the computed scattered intensity curve for a *trans* model fitted reasonably well the experimental diffraction pattern except in the low angle region. Attention is called to an infrared-Raman study by Hayden, *et al.*[239], which strongly support the *trans* configuration both in the gaseous and solid states. For the spectroscopic study the sample was obtained from the same commercial source as that used in the electron diffraction study. Clearly, further work is needed to establish the lowest energy configuration for this molecule. The interpretation of electron diffraction patterns often suffers from limitations in the method of analysis, especially for cases where there are large amplitude motions. In case of $CF_3N{=}NCF_3$, a careful analysis of the low frequencies (≈ 50 cm^{-1})[239] for the torsional motion of CF_3 about the N–CF_3 bond should be made. The determination of the configuration of this molecule from electron diffraction data depends mainly on the position of the fluorine atoms (*i.e.*, nonbonded fluorine pairs between the two CF_3 groups). Hence the mean square amplitudes of the F . . . F pairs, or the large amplitude analysis of the CF_3 groups should provide additional useful information.

Propenes and Acetaldehydes

In the fluoro-substituted propenes and acetaldehydes interesting features appear relative to the potential function for the methyl rotor. Values for the barrier height (V_3) for the threefold barriers of CX_3 groups in CX_3–CY=Z,

Table 25. The threefold barriers, V_3 in propene, acetaldehyde, and their fluoro derivatives (gas phase).

Molecules	V_3 (kcal/mole)	Refs.
CH_3–CH=CH_2	2.00	193)
CH_3–CF=CH_2	2.4 ± 0.1	195)
CH_3–CH=CHF (*trans*)	2.2 ± 0.1	53)
CH_3–CH=CHF (*cis*)	1.06 ± 0.05	52)
CH_3–CH=CF_2	1.25 ± 0.02	240)
CF_3–CF=CF_2	≈ 1.0	241)
CH_3–CH=O	1.16 ± 0.03	197)
CH_3–CF=O	1.04	55)
CF_3–CH=O	0.89 ± 0.08	242)
CF_3–CF=O	1.4 ± 0.2	243)

where X is H or F, Y is H or F, and Z is CH_2, CHF, CF_2 or O determined from microwave spectroscopy are given in Table 25. In each case at the minimum energy orientation one of the C–X bonds of CX_3 is located at an eclipsed position relative to the double bond. The striking effect of *cis* fluorination on the barrier height for propene is apparent. Scarzafava and Allen[244)] accounted for relative magnitudes of the barriers in propene and in its fluoroderivatives by allocating to various factors parts of the total energy calculated from an *ab initio* LCAO–MO analysis. They concluded that the *cis*-fluoro barrier was attractive dominant, while the barriers in propene and other fluoroderivatives were essentially dominated by repulsive interactions. For the *cis* isomer, ΔV_{ne} (nuclear-electronic attraction term) appears to be primarily responsible for stabilizing the eclipsed configuration. While no computed or measured equilibrium ratios for *cis-trans* isomers has been published for $CH_3CH{=}CHF$, it is interesting to note that the cis isomer of the chloro and bromo analogs are more stable, at least in solution[52, 222, 227)]. In 3-fluoropropene, $CH_2F{-}CH{=}CH_2$, the *cis* form is more stable than the *gauche* form by 166 ± 67 cal/mole.[29)]

Hexafluoropropene, $CF_3{-}CF{=}CF_2$, is of special interest since the electron diffraction data did not lead to a unique structure determination;[245)] both non-planar and planar

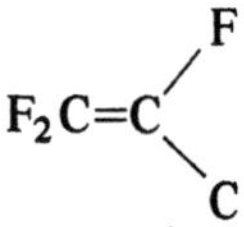

skeletons fitted the data. However, a recent microwave study by Jacob and Lide[241)] showed that the atoms are coplanar except for the two equivalent F atoms in the CF_3 groups. They could not determine the geometric parameters due to the absence of more than one F isotopic species, but their assumed structure fitted the observed moments of inertia. They also estimated a barrier height of approximately 1 kcal/mole for rotation of the CF_3 group and a torsional frequency of 30 ± 15 cm^{-1}. Such a low barrier and a low torsional oscillation of the CF_3 motion additionally complicates the interpretation of the diffraction patterns. An analysis of the diffraction data allowing for large amplitude motions will provide further information on the dynamic aspects and geometrical structure of this molecule using spectroscopic information.

1,3-Butadienes and Related Compounds

Rotational isomerism about the central C–C single bond in conjugated molecules, such

Table 26. The most stable conformations of 1,3-butadienes

Molecule	Most stable form (Φ)[1)]	Refs.
$CH_2{=}CH{-}CH{=}CH_2$	*trans* (180°)	143, 246)
$CH_2{=}CH{-}CX{=}CH_2$ (X = F, Cl, Br, I)	*trans* (180°)	249, 250)
$CH_2{=}CX{-}CX{=}CH_2$ (X = Cl, Br)	*trans* (180°)	232)
$CF_2{=}CH{-}CH{=}CF_2$	*trans* (180°)	251)
$CCl_2{=}CH{-}CCl{=}CH_2$	*gauche* (50°)	250)
$CBr_2{=}CH{-}CBr{=}CH_2$	*gauche* (50°)	250)
$CCl_2{=}CCl{-}CCl{=}CCl_2$	*gauche* (0–90°)	252)
$CF_2{=}CF{-}CF{=}CF_2$	*gauche* (47°)	103, 253)

[1)] Approximate dihedral angle, measured from the *cis*-planar form.

as butadiene presents an interesting structural feature. Thus, it is now established that the most stable form of 1,3-butadiene is s-*trans*.[143, 246] Aston, *et al.*[247] estimated the potential for rotation about the central bond from calorimetric data. They used a potential function with two minima, corresponding to the *trans* and *cis*-planar forms, with an energy difference of 2.3 kcal/mole. However, it is far from clear whether the second conformer is the *cis*-planar form or a nonplanar *gauche* form. Recent *ab initio* SCF calculations by Skancke, *et al.*[248] suggest that indeed the second stable form of this molecule (2.7 kcal/mole above *trans*) is a *gauche* conformation with a dihedral angle of about 40° away from the *cis*-planar form.

Halosubstituted 1,3-butadienes provide useful information on the configurational problem about the central single bond; the observed stable forms of halo-1,3-butadienes are listed in Table 26. On this basis the potential function for rotation about the central bond has three minima, corresponding to the *trans*-planar form and to the right and left *gauche* nonplanar forms (a dihedral angle of about 50° from the *cis*) as has been predicted by the SCF calculations for 1,3-butadiene. The relative stabilities of the two forms depend sensitively on the type of substitution (Table 26). It appears that 1,1,3-substitutions make the *gauche* minima lower than the *trans*.

Recently Hagen and Hedberg[254] studied several oxalyl halides by electron diffraction. Their data demonstrated the presence of *trans-gauche* conformers rather than *trans-cis* conformers. Similar questions regarding orientation about the C—C single bond in the conjugated molecules, biphenyl and the haloderivatives have been extensively studied by Bastiansen and co-workers.[255, 256]

Acetylacetones

The most common β-diketone, acetylacetone, exists in two tautomeric forms,

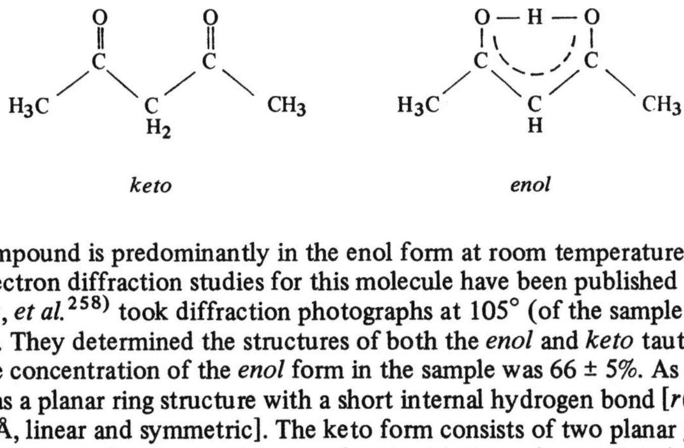

The compound is predominantly in the enol form at room temperature[257]. Two independent electron diffraction studies for this molecule have been published recently.[107, 258] Lowrey, *et al.*[258] took diffraction photographs at 105° (of the sample container and nozzle). They determined the structures of both the *enol* and *keto* tautomers and estimated that the concentration of the *enol* form in the sample was 66 ± 5%. As expected, the enol form has a planar ring structure with a short internal hydrogen bond [r(O–H–O) = 2.38 ± 0.02 Å, linear and symmetric]. The keto form consists of two planar groups, each incorporates the central carbon and an acetyl unit with their planes rotated relative to each other through a dihedral angle of 48.6 ± 4°. Lowrey, *et al.* reported that the bond distances in the two tautomers differ significantly. On the other hand, Andreassen, *et al.*[107] took their diffraction photographs with sample reservoir at room temperature where the enol form comprises all but a negligible fraction of the sample. For the enol the two structure determinations are in essential agreement; the largest discrepancy is in the O . . . H . . . O distance: 2.38 ± 0.02 Å (Lowrey, *et al.*[258]) and 2.52 ± 0.02 Å (Andreassen, *et al.*[107])

Structural changes associated with fluorine for hydrogen substitution in acetylacetone, trifluoro- and hexafluoro-acetylacetones were investigated by Andreassen, *et al.*[107, 259]

At room temperature only enol forms are present in the gas phase.[260] The structural parameters were determined on the basis of C_{2v} symmetry for the *enol* ring. These structures are illustrated in Fig. 15. The parameters for acetylacetone and acetone are averaged values of the independent electron diffraction studies. It thus appears that CF_3 for CH_3 substitution leads to shorter C $\overline{...}$ O distances in the enol tautomer, whereas in acetone

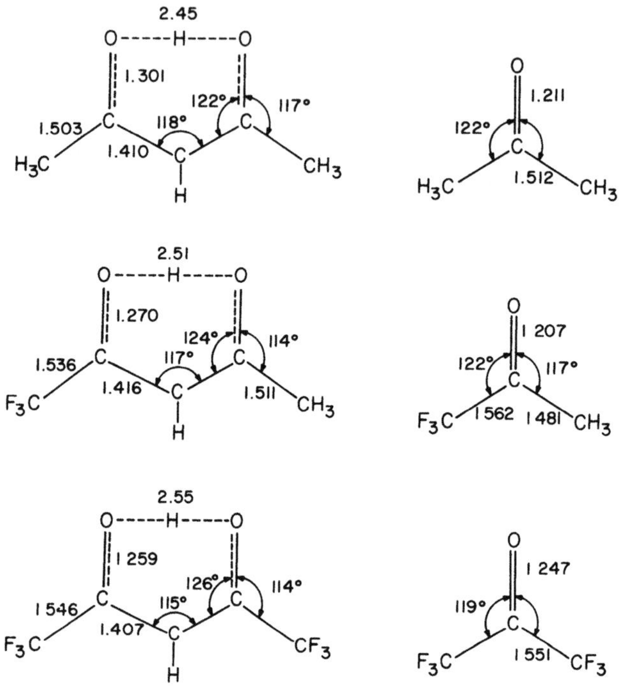

Fig. 15. Structures and the *enol* form for acetylacetone[107, 258], trifluoroacetylacetone[107], and hexafluoroacetylacetone[259] compared with corresponding structures in acetone[108, 133], trifluoroacetone[107], and hexafluoroacetone[108] (r_g values)

a single CF_3 leaves the C=O distance essentially unchanged while two CF_3's (hexafluoroacetone) induces a rather large extension of the C–O bond (~0.04 Å). The C–C distances external to the enol ring follow the same trend as in the acetones; the C–CF_3 bond is longer than the C–CH_3 bond. With respect to the fluorine substitution effect on acetylacetone, Gordon and Koob[261] calculated the structural parameters of the *enol* forms of acetylacetone and trifluoroacetylacetone using INDO-MO. They concluded that in the lowest energy form the H-bonded hydrogen in acetylacetone is symmetrically placed between the two oxygens, with an OHO angle of 152° and an O . . . O distance of 2.35 Å. The CF_3 for CH_3 substitution introduces only a slight asymmetry. They also predicted that CF_3 substitution results in an alternation of bond lengths and bond dissociation energies about the *enol* ring; the ring C–C bond on the CF_3 side is strengthened, the adjacent C–O bond is weakened, and the O---H bond is strengthened relative to acetylacetone. The opposite effect appears on the methyl side.

IV. Selected Inorganic Fluorides

The halides of group V elements comprise a series of inorganic compounds which have been studied systematically by combining electron diffraction and spectroscopic techniques. The internuclear distances and bond angles in MX_3 and YPX_3, where M is N, P, As, and Sb, X is F, Cl, Br, and I, and Y is O and S, are summarized in a review paper by Kuchitsu[2]. Trends observed in these molecules are well accounted for qualitatively by the valence-shell electron-pair repulsion theory (VSEPR)[74]

Hargittai, *et al.*[262–266] reported structural data on several sulphones and thionyl fluorides. They discussed the regularities in the molecular structures of X_2SO and $XYSO_2$, where X, Y are F, Cl, or CH_3. The VSEPR model also proved useful for this series. For instance, as the substituent electronegativity decreases from fluorine to methyl, the S–O bond lengthens and the magnitude of the O–S–O bond angle decreases. Related compounds, the polysulfuryl fluorides

$$(FO_2S\text{–}O\text{–}SO_2F \text{ and } FO_2S\text{–}O\text{–}SO_2\text{–}O\text{–}SO_2F)$$

were studied by Hencher and Bauer[267]. The observed S–O bonds and O–S–O angles in the $-SO_2F$ groups fall within the scope of the above structural regularities. The same trend applies to the system $X\text{–}NO_2$ (where X is CH_3[135], CF_3[116], F[268], and Cl[269]); the N–O bond distance lengthens and the O–N–O angle decreases as the ligand electronegativity decreases. Indeed, the VSEPR theory has proved valid and widely applicable for qualitative accounting of regularities in the structures of many inorganic compounds. Bartell[75] showed that the VSEPR model corresponds closely to a Hückel MO model, *neglecting* electron-electron electrostatic repulsions and ignoring nonbonded interactions. Thus, when the neglect of these factors becomes significant predictions based on the VSEPR model break down.

An accurate structure for BF_3 was determined by Kuchitsu and Konaka[270] who combined electron diffraction and spectroscopic data: r_g (B–F) = 1.313 ± 0.001 Å. The B–F distance in $F_3Si\text{–}SiF_2\text{–}BF_2$ by Chang, *et al.*[271] is 1.309 ± 0.002 Å, close to that in BF_3 while the B–F distance in $(SiH_3)_2N\text{–}BF_2$ by Robiette, *et al.*[272] is about 0.02 Å longer than in BF_3. For comparison, in crystalline $(CH_3)_3N{:}BF_3$, the B–F distance is 1.39 Å, (B–N) = 1.58 Å, and (N–C) = 1.50 Å.[273] The latter two values are equal, within rather large experimental error limits, to corresponding distances in $(CH_3)_3N{:}BH_3$.[274]

Hedberg, *et al.*,[275, 276] recently studied B_2F_4 and B_2Cl_4 by electron diffraction. An interesting difference appeared relative to their minimum energy conformations. They concluded that the potential hindering internal rotation in B_2Cl_4 is described by

$$V = \frac{1}{2} V_0 (1 - \cos 2\theta),$$

where V_0 = 1.85 ± 0.03$_5$ kcal/mole and θ is the dihedral angle between the two BCl_2 planes (measured from the coplanar configuration). In contrast to the non-planar C_{2v} ground state conformation of B_2Cl_4 in the gas phase, the lowest energy conformation of B_2F_4 is planar, D_{2h}, with a barrier V_0 of 0.4 kcal/mole[277]. The Raman spectra of B_2F_4 on both the gaseous and crystalline states support this conclusion[278]. At this stage the parent hydride, B_2H_4 has not been synthesized.

The fluorides of nitrogen and phosphorus provide interesting parallels and contrasts. The dinitrogen and diphosphorous compounds:

P_2H_4[279], H_2PPF_2[280], $P_2(CH_3)_4$[281], $P_2(CF_3)_4$[282], P_2F_4[282], and N_2H_4[283], $H_2N_2(CH_3)_2$[284], $N_2(CF_3)_4$[117], N_2F_4[285]

show similar trends in the lengths of the central bond for various substituents. The central bonds in P_2F_4 and N_2F_4 are 0.06 Å and 0.04 Å longer than those in P_2H_4 and N_2H_4, respectively. On the other hand, in $P_2(CF_3)_4$ and $N_2(CF_3)_4$ the central bonds are about 0.05 Å shorter than those in diphosphine and hydrazine. Bartell and Higginbotham[117], on the basis of a molecular orbital description, suggested that the short N–N bond in $N_2(CF_3)_4$ may be a consequence of enhanced π-bonding, which is facilitated by the nearly D_{2d} symmetry, imposed by steric forces. Concerning the lowest energy conformations of these compounds, N_2F_4, originally thought to exist only in the *gauche* form[286], has recently been found by Cardillo and Bauer[285] to be a mixture of ≈47% *gauche* and 53% *trans.* In contrast only single forms have been found for N_2H_4, $N_2(CF_3)_4$, and P_2H_4 (*gauche*); P_2F_4 and $P_2(CH_3)_4$ (*trans*). Except for N_2H_4[287], quantitative estimates of potential functions hindering rotation about the central bonds have not been made for these molecules.

The trend followed by the central bonds in HOOH[288], FOOF[289], CF_3OOCF_3[121], HSSH[290], CH_3SSCH_3[136], FSSF[291], and CF_3SSCF_3[123] contrasts with those in the diphosphines and hydrazines; both the O–O and S–S distances are shortened by fluorine substitution and lengthened by CF_3 substitution. The O–O and S–S distances in FOOF and FSSF are particularly unusual; they are respectively 0.26 Å and 0.16 Å shorter than in the corresponding hydrogen analogs, being close to bond lengths in O_2 and S_2. Several qualitative explanations have been proposed; they were formulated in the language of valence bond theory ($F^-O{=}O^+F$ and $F^-S{=}S^+F$ ionic structures), qualitative molecular orbital theory (Lipscomb[292]), and Linnett's theory (double quartet formulation of the octet[293]). Semi-empirical INDO calculations and *ab initio* MO computations by Pople, *et al.*[20,21] did not reproduce the observed magnitudes for the O–O distances in HOOH and FOOF.

In their stable conformations XOOX and XSSX frames are not coplanar; the dihedral angles are 90° (HSSH), 88° (FSSF), 84° (CH_3SSCH_3), 119.8° (HOOH), 87.5° (FOOF), and 59° (CF_3OOCF_3). The barriers hindering rotation in HOOH, estimated from an IR spectroscopic study by Hunt, *et al.*[294] are: V(cis) = 7.03 kcal/mole and V(trans) = 1.10 kcal/mole.

Replacement of all the H atoms by F's in H_2O and in NH_3 leads to a decrease in bond angles:

(r_0) 103.3° (F_2O) vs (r_0) 104.9° (H_2O); and
(r_e) 102.4° (NF_3)[12] vs (r_e) 106.7° (NH_3)[295]; see Table 1b.

These decrements in the bond angles are qualitatively explained by the VSEPR model. Specifically, the greater electronegativity of fluorine leads to a more compact N–F bond orbital, compared with the N–H orbital, and thus it occupies less space on the surface of the nitrogen atom. Because of repulsion from the lone electron pair, the angle between the less extended N–F bond orbitals becomes smaller than the angle between the more extended N–H bond orbitals. If these explanations are correct one may expect intermediate bond angles for HOF and NHF_2. However, the observed HOF (97.2°: r_s) and HNF (99.8°: r_s)[11] angles are much smaller than those in the corresponding perfluoro compounds. In addition, perfluorination of H_2S, PH_3, AsH_3, and SbH_3 causes the bond angles to be larger, about 6° for S[296, 297] and about 4° for P, As and Sb[298–303], in contrast with perfluoroination of H_2O and H_3N.

A systematic study of the series $(CH_3)_nPF_{5-n}$ [n = 0,1,2,3] has been recently completed by Bartell and co-workers[304]. The observed variations in the bond distance of these trigonal bipyramidal compounds are well correlated with the number of methyl substituents. In all cases the least electronegative ligands (or CH_3) occupy equatorial sites. The sterochemistry and trends in structure parameters are well accounted for by the VSEPR theory. Furthermore, the increase of the axial P–F bond lengths in this series correlates well with the increase in P–F amplitudes of vibration. The methyl groups essentially rotate freely.

The structure of the unexpected inert gas compound, XeF_6 has been investigated in great detail by Bartell, *et al.*[305, 306] The fluorine atoms execute large amplitude motions departing substantially from the idealized O_h structure. These motions are best described as three-dimensional pseudo rotations, similar to those presumed to be present in some five-member ring molecules[81, 307–309], in IF_7[310] and in ReF_7[311]. In contrast, XeF_2[312] and XeF_4[313] have rather rigid structures, linear and square planar, respectively.

V. Conclusions

The information assembled in this review of the gas phase molecular structures of fluorocarbons and related compounds were derived from electron diffraction patterns and microwave spectra. We placed primary emphasis on the geometrical effects of fluorine substitution (bond distances, bond angles and internal rotational parameters). Clearly other aspects of molecular structure, such as dipole moments, average bond dissociation energies, photoelectron spectra, etc. merit codification. At first glance, interatomic distances and bond angles call attention to features which appear to be strictly localized within molecules. It is evident, however, that such a local bond concept is not consistent with the totality of structural data on the level of precision currently available. Furthermore, it is also clear that the level of accuracy now generally attainable in the computation of molecular geometries is considerably below experimentally derived quantities. Regularities in the molecular structures of organic fluorides have been uncovered. For some structural types there are strong similarities and comparable trends among the fluorides of elements other than carbon. The substitution or environmental effects on molecular structures raise interesting questions regarding the nature of chemical bonds, and the contrasts between fluorine and hydrogen substituents provide an incisive focal points for theoreticians.

Acknowledgments
This study was supported by the National Science Foundation under Grant No. GP-34060. The authors wish to thank Professors M. Kimura and K. Hedberg for communicating their results prior to publication.

VI. References

1) Bauer, S. H.: Physical chemistry, an advanced treatise, Vol. 4, Chap. 1 and 14, (D. Henderson, ed.). New York: Academic Press. 1970.
2) Kuchitsu, K.: MTP International Review of Science, Vol. 2, Chap. 6, (G. Allen, ed.). London and Baltimore: Butterworths, University Park Press. 1973.
3) Hilderbrandt, R. L., and Bonham, R. A.: Ann. Rev. Phys. Chem. *22*, 279 (1971).
4) Karle, J.: Determination of organic structures by physical methods, Vol. 5, (F. C. Nachod and J. J. Zuckermann, eds.). New York: Academic Press 1972.
5) Bartell, L. S.: Physical methods in chemistry, 4th Edition, (A. Weissberger and B. W. Rossiter, eds.). New York: Interscience 1972.
6) For example: Proceedings of The Robert A. Welch Foundation Conferences on Chemical Research XVI. Theoretical Chemistry, (W. O. Milligan, ed.). Houston, Texas 1973.
7) Kuchitsu, K., and Cyvin, S. J.: Molecular vibrations and structure studies. Chap. 12, (S. J. Cyvin, ed.). Amsterdam: Elsevier 1972.
8) Sutton, L. E. (ed.): Tables of interatomic distances and configurations in molecules and ions. London: Chemical Society 1958, and Supplement 1965.
9) Herzberg, G.: Science *177*, 123 (1972).
10) Suchard, S. N.: Spectroscopic constants for selected hetronuclear diatomic molecules, Vols. I, II, and III. Aerospace Report No. TR-0074(4641)-6 (1974).
11) Lide, D. R., Jr.: J. Chem. Phys. *38*, 456 (1963).
12) Ootake, M., Matsumura, C., and Morino, Y.: J. Mol. Spectrosc. *28*, 325 (1968).
13) Takagi, K., and Oka, T.: J. Phys. Soc. Japan *18*, 1174 (1963).
14) Miller, R. F., and Curl, R. F., Jr.: J. Chem. Phys. *34*, 1847 (1961).
15) Carpenter, J. H.: J. Mol. Spectrosc. *50*, 182 (1974).
16) Brockway, L. O.: J. Phys. Chem. *41*, 185, 747 (1937).
17) Pauling, L.: The nature of the chemical bond, 3rd Edition, Ithaca, New York: Cornell University Press 1960.
18) Peters, D.: J. Chem. Phys. *38*, 561 (1963); Bent, H. A.: J. Chem. Phys. *33*, 304, 1258, 1259, 1260 (1960).
19) Hine, J.: J. Am. Chem. Soc. *85*, 3239 (1963).
20) Gordon, M. S., and Pople, J. A.: J. Chem. Phys. *49*, 4643 (1968).
21) Newton, M. D., Lathan, W. A., Hehre, W. J., and Pople, J. A.: J. Chem. Phys. *52*, 4064 (1970).
22) Bader, R. M.: J. Chem. Phys. *2*, 128 (1934), and *3*, 710 (1935).
23) Dewar, M. J. S., and Schmeising, H. N.: Tetrahedron *5*, 166 (1959) and *11*, 96 (1960).
24) Muller, N., and Pritchard, D. E.: J. Chem. Phys. *31*, 768, 471 (1959).
25) Thornton, C. G.: Diss. Abs. *14*, 604 (1953).
26) Costain, C. C.: J. Chem. Phys. *29*, 864 (1958).
27) Nygaard, L.: Spectrochim. Acta *22*, 1261 (1966).
28) Van Eijck, B. P., Van der Plaats, G., and Van Roon, P. H.: J. Mol. Structure *11*, 67 (1972).
29) Hirota, E.: J. Chem. Phys. *42*, 2071 (1965).
30) Hagen, K., and Hedberg, K.: J. Am. Chem. Soc. *95*, 8263 (1973).
31) Brunovoll, J.: Thesis, Trondheim (1962).
32) Van Schaick, E. J. M., Geise, H. J., Mijlhoff, F. C., and Renes, G.: J. Mol. Structure *16*, 23 (1973).
33) Andersen, P.: Acta Chem. Scand. *16*, 2337 (1962).
34) Kimura, M.: 5th Austin Symposium on Gas Phase Molecular Structure, Austin, Texas 1974.
35) Stolenvik, R., and Thom, E.: Acta Chem. Scand. *25*, 3205 (1971).
36) Lide, D. R., Jr., and Mann, D. E.: J. Chem. Phys. *29*, 914 (1958).
37) Schwendeman, R. H., and Jabob, G. D.: J. Chem. Phys. *36*, 1245 (1962).
38) Tobiason, F. L., and Schwendemann, R. H.: J. Chem. Phys. *40*, 1014 (1964).
39) Lide, D. R., Jr., and Jen, M.: J. Chem. Phys. *38*, 1504 (1963).
40) Flanagan, C., and Pierce, L.: J. Chem. Phys. *38*, 2963 (1963).
41) Schwendeman, R. H., and Tobiason, F. L.: J. Chem. Phys. *43*, 201 (1965).
42) Lide, D. R., Jr.: J. Chem. Phys. *33*, 1514 (1960).
43) Kuchitsu, K.: J. Chem. Phys. *49*, 4456 (1968).
44) Lide, D. R., Jr.: J. Chem. Phys. *33*, 1519 (1960).
45) Hilderbrandt, R. L., and Wieser, J. D.: J. Mol. Structure *15*, 27 (1973).
46) Chiang, J. F., Wilcox, C. F., Jr., and Bauer, S. H.: Tetrahedron *25*, 369 (1969).
47) Chiang, J. F., Wilcox, C. F., Jr., and Bauer, S. H.: J. Am. Chem. Soc. *90*, 3149 (1968).
48) Hilderbrandt, R. L., and Wieser, J. D.: J. Chem. Phys. *55*, 4648 (1971).
49) Schwendeman, R. H., Jacob, G. D., and Krigas, T. M.: J. Chem. Phys. *40*, 1022 (1964).

50) Carlos, J. L., Jr., Karl, R. R., Jr., and Bauer, S. H.: Faraday Trans. II *70*, 177 (1974).
51) Laurie, V. W.: J. Chem. Phys. *34*, 291 (1961).
52) Beaudet, R. A., and Wilson, E. B., Jr.: J. Chem. Phys. *37*, 1133 (1962).
53) Siegel, S.: J. Chem. Phys. *27*, 989 (1957).
54) Jones, M. E., Hedberg, K., and Schomaker, V.: J. Am. Chem. Soc. *77*, 5278 (1955).
55) Pierce, L., and Krisher, L. C.: J. Chem. Phys. *31*, 875 (1959).
56) Nygaard, L., Bojesen, I., Pedersen, T., and Rastrup-Andersen, J.: J. Mol. Structure *2*, 209 (1968).
57) Hatta, A., Hirose, C., and Kojima, K.: Bull. Chim. Soc. Japan *41*, 1088 (1968).
58) Nygaard, L., Hansen, E. R., Hansen, R. L., Rastrup-Andersen, J., and Sorensen, G. O.: Spectrochim. Acta *23A*, 2813 (1967).
59) Van Schaick, E. J. H., Geise, H. J., Mijlhoff, F. C., and Renes, G.: J. Mol. Structure *16*, 389 (1973).
60) Bauer, S. H., Katada, K., and Kimura, K.: Structural chemistry and molecular biology; pp. 653. San Francisco: W. H. Freeman and Co. 1968.
61) Schlupf, J., and Weber, A.: J. Raman Spectroscopy *1*, 3 (1973).
62) Doraiswamy, S., and Scharma, S. D.: Proc. Ind. Acad. Sci. *A68*, 53 (1968).
63) Almenningen, A., Bastiensen, O., Seip, R., and Seip, H. M.: Acta Chem. Scand. *18*, 2115 (1964).
64) Epiotis, N. D.: J. Am. Chem. Soc. *95*, 3087 (1973).
65) Tyler, J. K., and Sheridan, J.: Trans. Faraday Soc. *59*, 2661 (1963).
66) Karakida, K., Fukuyama, T., and Kuchitsu, K.: Bull. Chem. Soc. Japan *47*, 299 (1974).
67) Lafferty, W. J., Plyler, E. K., and Tidwell, E. D.: J. Chem. Phys. *37*, 1981 (1962).
68) Laurie, V. W., and Herchbach, D. R.: J. Chem. Phys. *37*, 1687 (1962).
69) Fukuyama, T., Kuchitsu, K., and Morino, Y.: Bull. Chem. Soc. Japan *42*, 379 (1969).
70) Fukuyama, T., and Kuchitsu, K.: J. Mol. Structure *5*, 131 (1970).
71) Laurie, V. W., and Pence, D. T.: J. Chem. Phys. *37*, 2995 (1962).
72) Laurie, V. W., and Pence, D. T.: J. Chem. Phys. *38*, 2693 (1963).
73) Kristoff, J. J., Jr.: Diss. Abs. *18*, 1997 (1958); Livingston, R. L., and Lyon, D. H.; J. Chem. Phys. *24*, 1283 (1956).
74) Gillespie, R. J.: J. Chem. Educ. *40*, 295 (1963) and *47*, 18 (1970).
75) Bartell, L. S.: J. Chem. Educ. *45*, 754 (1968).
76) Chiang, J. F., and Bernett, W. A.: Tetrahedron *27*, 975 (1971).
77) Chang, C. H., Porter, R. F., and Bauer, S. H.: J. Mol. Structure *7*, 89 (1971).
78) Hjortaas, K. E., and Stromme, K. O.: Acta Chem. Scand. *22*, 2965 (1968).
79) Bastiansen, O., Fritsch, F. N., and Hedberg, K.: Acta Cryst. *17*, 538 (1964).
80) Almenningen, A., Bastiansen, O., and Skancke, P. N.: Acta Chem. Scand. *15*, 711 (1961).
81) Adams, W. J., Geise, H. J., and Bartell, L. S.: J. Am. Chem. Soc. *92*, 5013 (1970).
82) Bastiansen, O., Fernholt, L., Seip, H. M., Kambara, H., and Kuchitsu, K.: J. Mol. Structure *18*, 163 (1973).
83) Walsh, A. D.: Nature *159*, 165, 712 (1947); Trans. Faraday Soc. *45*, 179 (1949).
84) Maciel, G. E., McIver, J. W., Jr., Ostlund, N. S., and Pople, J. A.: J. Am. Chem. Soc. *92*, 1 (1970).
85) Juan, C., and Gutowsky, H. S.: J. Chem. Phys. *37*, 2198 (1962).
86) Karplus, M., and Grant, D. M.: Proc. Natl. Acad. Sci. U. S. *45*, 1269 (1959).
87) Wiberg, K. B., Lampman, G. M., Ciula, R. P., Connor, D. S., Schertler, P., and Lavanish, J.: Tetrahedron *21*, 2749 (1965).
88) Lynden-Bell, R. M., and Sheppard, N.: Proc. Roy. Soc. *A269*, 385 (1962); Graham, D. M., and Holloway, C. E.: Can. J. Chem. *41*, 2114 (1963).
89) Hutton, H. M., Reynolds, W. F., and Schaefer, T.: Can. J. Chem. *40*, 1758 (1962).
90) Weigert, F. J., and Roberts, J. D.: J. Am. Chem. Soc. *93*, 2361 (1971).
91) Patel, D. J., Howden, M. E. H., and Roberts, J. D.: J. Am. Chem. Soc. *85*, 3218 (1963).
92) Hencher, J. L., and Bauer, S. H.: J. Am. Chem. Soc. *89*, 5527 (1967).
93) Chang, C. H., and Bauer, S. H.: J. Phys. Chem. *75*, 1685 (1971).
94) Ghosh, S. N., Trumbarulo. R., and Gordy, W.: J. Chem. Phys. *20*, 605 (1952).
95) Schwendeman, R. H.: Diss. Abs. *18*, 1645 (1958).
96) Edgell, W. F., Miller, G. B., and Amy, J. W.: J. Am. Chem. Soc. *79*, 2391 (1957).
97) Tokue, I., Fukuyama, T., and Kuchitsu, K.: J. Mol. Structure *17*, 207 (1973).
98) Shoolery, J. N., Shulman, R. G., Sheehan, W. F., Jr., Schomaker, V., and Yost, D. M.: J. Chem. Phys. *19*, 1364 (1951).
99) Anderson, R. E.: Diss. Abs. *19*, 50, 51 (1958).
100) Thomas, L. F., Heeks, J. S., and Sheridan, J.: Z. Electrochim. *61*, 935 (1957); Sheridan, J., and Gordy, W.: J. Chem. Phys. *20*, 591 (1952).
101) Chang, C. H., Porter, R. F., and Bauer, S. H.: J. Am. Chem. Soc. *92*, 5313 (1970).
102) Laurie, V. W.: J. Chem. Phys. *30*, 1101 (1959).

103) Chang, C. H., Andreassen, A. L., and Bauer, S. H.: J. Org. Chem. *36*, 920 (1971).
104) Kveseth, K., Seip, H. M., and Stolenvik, R.: Acta Chem. Scand. *25*, 2975 (1971).
105) Yokozeki, A.: 5th Austin Symposium on Gas Phase Molecular Structure, Abstract W5, Austin, Texas 1974.
106) Tipton, A. B., Britt, C. O., and Boggs, J. E.: J. Chem. Phys. *46*, 1606 (1967).
107) Andreassen, A. L., and Bauer, S. H.: J. Mol. Structure *12*, 381 (1972).
108) Hilderbrandt, R. L., Andreassen, A. L., and Bauer, S. H.: J. Phys. Chem. *74*, 1586 (1970).
109) Marsden, C. J., and Sheldrick, G. M.: J. Mol. Structure *10*, 405 (1971).
110) Bartell, L. S., and Brockway, L. O.: J. Chem. Phys. *23*, 1860 (1955).
111) Andreassen, A. L., and Bauer, S. H.: J. Chem. Phys. *56*, 3802 (1972).
112) Marsden, C. J., and Sheldrick, G. M.: J. Mol. Structure *10*, 419 (1971).
113) Gallaher, K. L., Yokozeki, A., and Bauer, S. H.: J. Chem. Phys. (in press) (1974).
114) Bauer, S. H., and Andreassen, A. L.: J. Phys. Chem. *76*, 3099 (1972).
115) Davis, M. I., Boggs, J. E., Coffey, D., Jr., and Hanson, H. P.: J. Phys. Chem. *69*, 3727 (1965).
116) Karle, I. L., and Karle, J.: J. Chem Phys. *36*, 1969 (1962).
117) Bartell, L. S., and Higginbotham, H. K.: Inorg. Chem. *4*, 1346 (1965).
118) Glidowell, C., Rankin, D. W. H., Robiette, A. G., Sheldrick, G. M., and Williamson, M.: J. Chem. Soc. (A) 478 (1971).
119) Diodati, F. P., and Bartell, L. S.: J. Mol. Structure *8*, 395 (1971).
120) Hoffman, C. W. W., and Livingston, R. L.: J. Chem. Phys. *21*, 565 (1953).
121) Hilderbrandt, R. L., and Bauer, S. H.: unpublished paper (1970); S. H. Bauer, 3rd Austin Symposium on Gas Phase Molecular Structure, Austin, Texas 1970.
122) Kimura, K., and Kubo, M.: J. Chem. Phys. *30*, 151 (1959).
123) Bowen, H. J. M.: Trans. Faraday Soc. *50,* 452 (1954).
124) Pierce, L., and Hayashi, M.: J. Chem. Phys. *35*, 479 (1961).
125) Goldish, E., Hedberg, K., Marsh, R. E., and Schomaker, V.: J. Am. Chem. Soc. *77*, 2948 (1955).
126) Livingston, R. L., and Vanghan, G.: J. Am. Chem. Soc. *78,* 4866 (1956).
127) Beagley, B., and Hewitt, T. G.: Trans. Faraday Soc. *64*, 2561 (1968).
128) Bowen, H. J. M.: Trans. Faraday Soc. *50*, 463 (1954).
129) Bartell, L. S., and Brockway, L. O.: J. Chem. Phys. *32*, 512 (1960).
130) Springall, H. D., and Brockway, L. O.: J. Am. Chem. Soc. *60*, 996 (1938).
131) Bartell, L. S., and Bonham, R. A.: J. Chem. Phys. *32*, 824 (1960).
132) Tanimoto, M., Kuchitsu, K., and Morino, Y.: Bull. Chem. Soc. Japan *42*, 2519 (1969).
133) Kato, C., Konaka, S., Iijima, T., and Kimura, M.: Bull. Chem. Soc. Japan *42*, 2148 (1969); Iijima, T., and Tsuchiya, S.: J. Mol. Spectrosc. *44*, 88 (1972).
134) Coffey, D., Jr., Britt, C. O., and Boggs, J. E.: J. Chem. Phys. *49*, 591 (1968).
135) Cox, A. P., and Waring, S.: Trans. Faraday Soc. 1060 (1972).
136) Beagley, B., and McAloon, K. T.: Trans. Faraday Soc. *67*, 3216 (1971).
137) D'Antonio, P., George, C., Lowrey, A. H., and Karle, J.: J. Chem. Phys. *55*, 1071 (1971).
138) Wang, I. Y. M., Britt, C. O., Corwley, H., and Boggs, J. E.: J. Chem. Phys. *48*, 812 (1968).
139) Kojima, T., Breig, E., and Lin, C. C.: J. Chem. Phys. *35*, 2139 (1961).
140) Bartell, L. S.: J. Chem. Phys. *32*, 832 (1960).
141) Pople, J. A., and Gordon, M.: J. Am. Chem. Soc. *89*, 4253 (1967).
142) Egger, K. W., and Cocks, A. T.: Helv. Chim. Acta *56*, 1516, 1537 (1973).
143) Kuchitsu, K., Fukuyama, T., and Morino, Y.: J. Mol. Structure *1*, 463 (1968), *4*, 41 (1969).
144) Radom, L., Hehre, W. J., and Pople, J. A.: J. Am. Chem. Soc. *93,* 289 (1971).
145) Bak, B., Kierkegaard, C., Pappas, J., and Skaarup, S.: Acta Chem. Scand. *27*, 363 (1973).
146) Almenningen, A., Bastiansen, O., Motzfeldt, T.: Acta Chem. Scand. *23*, 2848 (1969).
147) Sugie, M., Fukuyama, T., and Kuchitsu, K.: J. Mol. Structure *14*, 333 (1972).
148) Kato, C., Konaka, S., Iijima, T., and Kimura, M.: Bull. Chem. Soc. Japan *42*, 2148 (1969).
149) Bartell, L. S., Kuchitsu, K., and DeNeui, R. J.: J. Chem. Phys. *35*, 1211 (1961).
150) Hirota, E., Sugisaki, R., Nielsen, C. J., and Sorensen, G. O.: J. Mol. Spectrosc. *49*, 251 (1974); For the r_g structure, see Kitano, M., and Kuchitsu, K.: Bull. Chem. Soc. Japan, *47*, 67 (1974).
151) Nishikawa, T., Itoh, T., and Shimoda, K.: J. Chem. Phys. *23*, 1735 (1955); Lide, D. R., Jr.; J. Chem. Phys. *27*, 343 (1957).
152) Kitano, M., and Kuchitsu, K.: Bull. Chem. Soc. Japan *46*, 3048 (1973).
153) Derissen, J. L.: J. Mol. Structure *7*, 67 (1971).
154) Sugie, M., and Kuchitsu, K.: J. Mol. Structure *20*, 437 (1974).
155) Tsuchiya, S., and Kimura, M.: Bull. Chem. Soc. Japan *45*, 736 (1972).
156) Higginbotham, H. K., and Bartell, L. S.: J. Chem. Phys. *42*, 1131 (1965).
157) Iijima, T.: Bull. Chem. Soc. Japan *45*, 1291 (1972).

158) Haase, B., Haase, J., and Zeil, W.: Z. Naturforsh *22a*, 1646 (1967).
159) Swick, D. A., and Karle, I. L.: J. Chem. Phys. *23*, 1499 (1955).
160) Pappas, J. A.: J. Mol. Structure *20*, 197 (1974).
161) Kuchitsu, K.: J. Chem. Phys. *44*, 906 (1966); Bartell, L. S., Roth, E. A., Hollowell, C. D., Kuchitsu, K. K., and Young, J. E., Jr.: J. Chem. Phys. *42*, 2683 (1965).
162) Ivey, R. C., and Davis, M. I.: J. Chem. Phys. *57*, 1909 (1972).
163) Davis, M. I., Kappler, H. A., and Cowan, D. J.: J. Phys. Chem. *68*, 2005 (1964).
164) Davis, M. I., and Hanson, H. P.: J. Phys. Chem. *69*, 4091 (1965).
165) Strand, T. G.: Acta Chem. Scand. *21*, 2111 (1967).
166) Strand, T. G.: Acta Chem. Scand. *21*, 1033 (1967).
167) Carlos, J. L., and Bauer, S. H.: Trans. Faraday Soc. *70*, 171 (1974).
168) Eisma, S. W., Altona, C., Geise, M. J., Mijlhoff, F. C., and Renes, G. H.: J. Mol. Structure *20*, 251 (1974).
169) Hope, H.: Acta Chem. Scand. *22*, 1057 (1968).
170) Almenningen, A., Anfinsen, I. M., and Haaland, A.: Acta Chem. Scand *24*, 43 (1970).
171) Morino, Y., Kuchitsu, K., Hori, Y., and Tanimoto, M.: Bull. Chem. Soc. Japan *41*, 2349 (1968).
172) Tanimoto, M., Kuchitsu, K., and Morino, Y.: Bull. Chem. Soc. Japan *44*, 386 (1971).
173) Zeil, W. O., Bodenseh, H. K., Gegenheimer, R., and Haase, J.: Bull. Am. Phys. Soc. Ser. II *13*, 832 (1968).
174) Pierce, L., and Dobyns, V.: J. Am. Chem. Soc. *84*, 2651 (1962).
175) Bak, B., Led, J. J., Nygaard, L., Rastrup-Andersen, J., and Sorensen, G. O.: J. Mol. Structure *3*, 369 (1969).
176) Davis, M. I., and Muecke, T. W.: J. Phys. Chem. *74*, 1104 (1970).
177) Brooks, W. V., Cyvin, B. N., Cyvin, S. J., Kvande, P. C., and Meirngeth, E.: Acta Chem. Scand. *17*, 345 (1963).
178) Strand, T. G., and Cox, H. L., Jr.: J. Chem. Phys. *44*, 2426 (1966).
179) Gallaher, K. L., and Bauer, S. H.: J. Phys. Chem. (in Press).
180) Wilson, E. B., Jr.: Advances in chemical physics, Vol. 2, (I. Prigogine, p. ed.), New York: Interscience 1959.
181) Lowe, J. P.: Progr. Phys. Org. Chem. *6*, 1 (1968).
182) Mizushima, S.: Structure of Molecules and Internal Rotation, Academic Press New York 1954.
183) Morino, Y., and Hirota, E.: Annu. Rev. Phys. Chem. *20*, 139 (1969).
184) During, J. R., Craven, S. M., and Harris, W. C.: Vibrational spectra and structure, Vol. 1, (J. R. During, ed.) p. 73. New York: Marcel Dekker, Inc. 1972.
185) Flygare, W. H.: Ann. Rev. Phys. Chem. *18*, 325 (1967).
186) Lowe, J. P.: Science *179*, 527 (1973).
187) For example: Radom, L., Hehre, W. J., and Pople, J. A.: J. Am. Chem. Soc. *94*, 2371 (1972).
188) Weiss, S., and Leroi, G. E.: J. Chem. Phys. *48*, 962 (1968).
189) Harschbach, D. R.: J. Chem. Phys. *25*, 358 (1956).
190) Wulff, C. A.: J. Chem. Phys. *39*, 1227 (1963).
191) Danti, A., and Wood, J. L.: J. Chem. Phys. *30*, 582 (1959).
192) Lowe, J. P., and Parr, R. G.: J. Chem. Phys. *44*, 3001 (1966).
193) Lowe, J. P.: J. Chem. Phys. *45*, 3059 (1966).
194) Fately, W. G., and Miller, F. A.: Spectrochim. Acta *19*, 611 (1963).
195) Pierce, L, and O'Reilly, J. M.: J. Mol. Spectrosc. *3*, 536 (1959); and Ref. 194.
196) Unland, M. L., Weiss, V. W., and Flygare, W. H.: J. Chem. Phys. *42*, 2138 (1965).
197) Verdier, P. H., and Wilson, E. B., Jr.: J. Chem. Phys. *29*, 340 (1958); Fately, W. G., and Miller, F. A.: Spectrochim. Acta *17*, 857 (1961).
198) Sinnott, K. M.: J. Chem. Phys. *34*, 851 (1961).
199) Krisher, L. C.: J. Chem. Phys. *33*, 1237 (1960).
200) Westheimer, F. H.: Steric effects in organic chemistry, (M. S. Newman, ed.), New York: Wiley 1956; Hendrickson, J. B.: J. Am. Chem. Soc. *83*, 4537 (1961).
201) Abraham, R. J., and Parry, K.: J. Chem. Soc. *B1970*, 539.
202) Scott, R. A., and Scheraga, H. A.: J. Chem. Phys. *42*, 2209 (1965).
203) Gordon, M. S.: J. Am. Chem. Soc. *91*, 3122 (1969).
204) Klaboe, P., and Nielsen, J. R.: J. Chem. Phys. *33*, 1764 (1960).
205) Ainsworth, J., and Karle, J.: J. Chem. Phys. *20*, 425 (1952).
206) Mizushima, S., Morino, Y., Watanabe, I., Shimanouchi, T., and Yamaguchi, S.: J. Chem. Phys. *17*, 591 (1949).
207) Bastiansen, O.: Acta Chem. Scand. *3*, 415 (1949).
208) Yokozeki, A., and Kuchitsu, K.: Bull. Chem. Soc. Japan *44*, 2922 (1971).

209) Kuchitsu, K.: Bull. Chem. Soc. Japan *32*, 748 (1959).
210) Fitzgerald, W. E., and Janz, G. J.: J. Mol. Spectrosc. *1*, 49 (1967).
211) Hirota, E.: J. Chem. Phys. *37*, 283 (1962).
212) Morino, Y., and Kuchitsu, K.: J. Chem. Phys. *28*, 175 (1958).
213) Igarashi, M., and Yamada, M.: Bull. Chem. Soc. Japan *26*, 871 (1956).
214) Almenningen, A., Bastiansen, O., Fernholt, L., and Hedberg, K.: Acta Chem. Scand. *25*, 1946 (1971).
215) Mizushima, S., Shimanouchi, T., Miyazawa, T., Abe, K., and Yasume, M.: J. Chem. Phys. *19*, 1477 (1951).
216) Giacomo, A. D., and Smyth, C. P.: J. Am. Chem. Soc. *77*, 1361 (1955).
217) Powling, J., and Bernstein, H. J.: J. Am. Chem. Soc. *73*, 1815 (1951).
218) Radom, L., Lathan, W. A., Hehre, J., and Pople, J. A.: J. Am. Chem. Soc. *95*, 693 (1973).
219) Bulthuis, J., Van der Berg, J., and Maclean, C.: J. Mol. Structure *16*, 11 (1973).
220) Pachler, K. G. R., and Tollenaere, J. P.: J. Mol. Structure *8*, 83 (1971).
221) Abraham, R. J., and Kemp, R. H.: J. Chem. Soc. *B1971,* 1240.
222) Viehe, H. G., Franchemont, E., Reenstein, M., and Valange, P.: Chem. Ber. *93*, 1697 (1960).
223) Demiel, A.: J. Org. Chem. *27*, 3500 (1962).
224) Wood, R. E., and Stevenson, D. P.: J. Am. Chem. Soc. *63*, 1650 (1941); Pitzer, K. S., and Hollenberg, J. L.: J. Am. Chem. Soc. *76*, 1493 (1954).
225) Viehe, H. G., and Franchimant, E.: Chem. Ber. *96*, 3153 (1963).
226) Crump, J. W.: J. Org. Chem. *28*, 953 (1963).
227) Harwell, K. E., and Hatch, L. E.: J. Am. Chem. Soc. *77*, 1682 (1955).
228) Fink, W. H., and Allen, L. C.: J. Chem. Phys. *46*, 2261, 2276 (1967).
229) Allen, L. C.: Chem. Phys. Lett. *2*, 597 (1968).
230) Liberles, A., Greeberg, A., and Eilers, J. E.: J. Chem. Educ. *50*, 676 (1973).
231) Trombetti, A., and Johns, J. W. C.: cited in J. W. Nibler and V. E. Bondybey, J. Chem. Phys. *60*, 1307 (1974).
232) Neisess, J. A.: Diss. Abs. Int. *32*, 2100B (1971).
233) Almenningen, A., Anfinsen, I. M., and Haaland, A.: Acta Chem. Scand. *24*, 1230 (1970).
234) Bohn, R. K., and Bauer, S. H.: Inorg. Chem. *6*, 309 (1967).
235) Kuczkowski, R. L., and Wilson, E. B., Jr.: J. Chem. Phys. *39*, 1030 (1963).
236) DeLange, J. J., Robertson, J. M., and Woodward, I.: Proc. Roy. Soc. Ser. A *171*, 398 (1939); Hampson, G. C., and Robertson, J. M.: J. Chem. Soc. *1941*, 409.
237) Armstrong, C. T., and Marantz, S.: J. Chem. Phys. *38*, 169 (1963).
238) Kaufman, J. J., Burnelle, L. A., and Haman, J. R.: Advances in chemistry series, American chemical society 1967, cited in Ref. 234.
239) Hayden, R. A., Tuazon, E. C., and Fateley, W. G.: J. Mol. Structure, *16*, 35 (1973).
240) Weiss, V. W., Beak, P., and Flygare, W. H.: J. Chem. Phys. *46*, 981 (1967).
241) Jacob, E. J., and Lide, D. R., Jr.: J. Chem. Phys. *59*, 5877 (1973).
242) Woods, R. C.: J. Chem. Phys. *46*, 4789 (1967).
243) Loos, K. R., and Lord, R. C.: Spectrochim. Acta *21*, 119 (1965).
244) Scarzafava, E., and Allen, L. C.: J. Am. Chem. Soc. *93*, 311 (1971).
245) Bauer, S. H., and Chang, C. H.: Abstract of papers of 161st meeting of the American Chemical Society, Los Angeles, California, 28 March 1971, Phys. 14; Buck, F. A. M., and Livingston, R. L.: J. Chem. Phys. *18*, 570 (1950).
246) Haugen, W., and Traetteberg, M.: Selected topics in structure chemistry, (P. Andersen, O. Bastiansen and S. Furberg, eds.), P. 113, Oslo: Universitetsforlaget 1967.
247) Aston, J. G., Szasz, G., Wooley, H. W., and Brickwedde, F. G.: J. Chem. Phys. *14*, 67 (1946).
248) Skancke, P. N., and Boggs, J. E.: J. Mol. Structure *16*, 179 (1973).
249) Lide, D. R., Jr.: J. Chem. Phys. *37*, 2074 (1962).
250) Bother-by, A. A., and Jung, D.: J. Am. Chem. Soc. *90*, 2342 (1968).
251) Beaudet, R. A.; J. Am. Chem. Soc. *87*, 1390 (1965).
252) Szasz, G., and Sheppard, N.: Trans. Faraday Soc. *49*, 358 (1953).
253) Robin, M., and Brundle, C. R.: J. Am. Chem. Soc. *92*, 5550 (1970).
254) Hagen, K., and Hedberg, K.: J. Am. Chem. Soc. *95*, 1003, 4796 (1973).
255) Bastiansen, O., and Traetteberg, M.: Tetrahedron *17*, 147 (1962).
256) Almenningen, A., Hartman, A. O., and Seip, H. M.: Acta Chem. Scand. *22*, 1013 (1968).
257) Gould, E. S.: Mechanism and structure in organic chemistry, p. 376 New York, Holt, Rinehart and Winston 1959; Burdett, J. L., and Rogers, M. T.: J. Am. Chem. Soc. *86*, 2105 (1964); Funke, E. and Mecke, R.: Hydrogen bonding, London: Pergamon Press 1959.
258) Lowrey, A. H., George, C., D'Antonio, P., and Karle, J.: J. Am. Chem. Soc. *93*, 6399 (1971).

259) Andreassen, A. L., Zebelman, D., and Bauer, S. H.: J. Am. Chem. Soc. *93*, 1148 (1971).
260) Conant, J. B., and Thompson, A. F., Jr.: J. Am. Chem. Soc. *54*, 4039 (1932); Schreck, R.: ibid. *71*, 1881 (1949); and Ref. 257.
261) Gordon, M. S., and Koob, R. D.: J. Am. Chem. Soc. *95*, 5863 (1973).
262) Hargittai, I., and Hargittai, M.: J. Mol. Structure *15*, 399 (1973).
263) Hargittai, M., and Hargittai, I.: J. Chem. Phys. *59*, 2513 (1973).
264) Hargittai, I.: Acta Chim. (Budapest) *60*, 231 (1969).
265) Hargittai, M., and Hargittai, I.: J. Mol. Structure *20*, 283 (1974).
266) Hargittai, I., and Mijlhoff, F. C.: J. Mol. Structure *16*, 69 (1973).
267) Hencher, J. S., and Bauer, S. H.: Can. J. Chem. *51*, 2047 (1973).
268) Legon, A. C., and Millen, D. J.: J. Chem. Soc., A 1736 (1968).
269) Millen, D. J., and Sinnott, K. M.: J. Chem. Soc. 350 (1958).
270) Kuchitsu, K., and Konaka, S.: J. Chem. Phys. *45*, 4342 (1966).
271) Chang, C. H., Porter, R. F., and Bauer, S. H.: J. Phys. Chem. *74*, 1363 (1970).
272) Robiette, A. G., Sheldrick, G. M., and Sheldrick, W. S.: J. Mol. Structure *5*, 423 (1970).
273) Geller, S., and Hoard, J. L.: Acta Cryst. *4*, 399 (1951).
274) Bauer, S. H.: J. Am. Chem. Soc. *59*, 1804 (1937).
275) Patton, J. V., and Hedberg, K.: Bull. Am. Phys. Soc. *13*, 831 (1968).
276) Ryan, R. R., and Hedberg, K.: J. Chem. Phys. *50*, 4986 (1969).
277) Danielson, D. D., and Hedberg, K.: (private communication), (1974); also cited in Ref. 278.
278) During, J. R., Thompson, J. W., Witt, J. D., and Odom, J. D.: J. Chem. Phys. *58*, 5339 (1973).
279) Beagley, B., Conrad, A. R., Freeman, J. M., Monaghan, J. J., and Norton, B. G.: J. Mol. Structure *11*, *371 (1972)*.
280) Kuczkowski, R. L., Schiller, H. W., and Rudolph, R. W.: Inorg. Chem. *10*, 2505 (1971).
281) McAdam, A., Beagley, B., and Hewitt, T. G.: Trans. Faraday Soc. *66*, 2732 (1970).
282) Hodges, L., and Bartell, L. S.: 4th Austin Symposium on Gas Phase Molecular Structure, Abstract M17, Austin, Texas 1972.
283) Morino, Y., Iijima, T., and Murata, Y.: Bull. Chem. Soc. Japan *33*, 46 (1960).
284) Beamer, W.: J. Am. Chem. Soc. *70*, 2979 (1948).
285) Cardillo, M. J., and Bauer, S. H.: Inorg. Chem. *8*, 2086 (1969).
286) Bohn, R. K., and Bauer, S. H.: Inorg. Chem. *6*, 304 (1967).
287) Kasuya, T., and Kojima, T.: J. Phys. Soc. Japan *18*, 364 (1963).
288) Redington, R. L., Olson, W. B., and Cross, P. C.: J. Chem. Phys. *36*, 1311 (1962).
289) Jackson, R. H.: J. Chem. Soc. *1962*, 4585.
290) Stevenson, D. P., and Beach, J. Y.: J. Am. Chem. Soc. *60*, 2872 (1938).
291) Kuczkowski, R. L.: J. Am. Chem. Soc. *86*, 3617 (1964).
292) Lipscomb, W. N.: cited in Ref. 289.
293) Linnett, J. W.: J. Chem. Soc. *1963*, 4663.
294) Hunt, R. H., Leacock, R. A., Peters, C. W., and Hecht, K. T.: J. Chem. Phys. *42*, 1931 (1965).
295) Kuchitsu, K., Guillory, J. P., and Bartell, L. S.: J. Chem. Phys. *49*, 2488 (1968).
296) Allen, H. C., and Plyler, E. K.: J. Chem. Phys. *25*, 1132 (1956).
297) Johnson, D. R., and Powell, F. X.: Science *164*, 950 (1969).
298) Burrus, C. A., Jache, A. W., and Gordy, W.: Phys. Rev. *95*, 700 (1954); Bartell, L. S., and Hirst, R. C.: J. Chem. Phys. *31*, 449 (1959).
299) Morino, Y., Kuchitsu, K., and Moritani, T.: Inorg. Chem. *8*, 867 (1969).
300) Blevins, G. S., Jache, A. W., and Gordy, W.: Phys. Rev. *97*, 684 (1955).
301) Clippard, F. B., Jr., and Bartell, L. S.: Inorg. Chem. *9*, 805 (1970); Konaka, S.: Bull. Chem. Soc. Japan *43*, 3107 (1970).
302) Jache, A. W., Blevins, G. S., and Gordy, W.: Phys. Rev. *97*, 680 (1955).
303) Ukaji, T., and Uchimura, H.: (1969) unpublished data cited in Ref. 2.
304) Yow, H., and Bartell, L. S.: J. Mol. Structure *15*, 209 (1973), and references cited therein.
305) Bartell, L. S.: J. Chem. Phys. *46*, 4530 (1967).
306) Bartell, L. S., and Gavin, R. M., Jr.: J. Chem. Phys. *48*, 2466 (1968).
307) Almenningen, A., Seip, H. M., and Willadsen, T.: Acta Chem. Scand. *23*, 2748 (1969).
308) Geise, H. J., Adams, W. J., and Bartell, L. S.: Tetrahedron *25*, 3045 (1969).
309) Engerholm G. G., Luntz, A. C., Gwinn, W. D., and Harris, D. O.: J. Chem. Phys. *50*, 2446 (1969).
310) Adams, W. J., Thompson, H. B., and Bartell, L. S.: J. Chem. Phys. *53*, 4040 (1970).
311) Jacob, E. J., and Bartell, L. S.: J. Chem. Phys. *53*, 2235 (1970).
312) Reichman, S., and Schreiner, F.: J. Chem. Phys. *51*, 2355 (1969).
313) Bohn, R. K., Katada, K., Marting, J. V., and Bauer, S. H.: Noble gas compounds, p. 238, Chicago University Press 1963.

Received June 18, 1974

GPSR Compliance

The European Union's (EU) General Product Safety Regulation (GPSR) is a set of rules that requires consumer products to be safe and our obligations to ensure this.

If you have any concerns about our products, you can contact us on ProductSafety@springernature.com

In case Publisher is established outside the EU, the EU authorized representative is:

Springer Nature Customer Service Center GmbH
Europaplatz 3
69115 Heidelberg, Germany

Batch number: 10620881

Printed by Printforce, the Netherlands